Lecture Notes in Networks and Systems 1637

The series "Lecture Notes in Networks and Systems" publishes the latest developments in Networks and Systems—quickly, informally and with high quality. Original research reported in proceedings and post-proceedings represents the core of LNNS.

Volumes published in LNNS embrace all aspects and subfields of, as well as new challenges in, Networks and Systems.

The series contains proceedings and edited volumes in systems and networks, spanning the areas of Cyber-Physical Systems, Autonomous Systems, Sensor Networks, Control Systems, Energy Systems, Automotive Systems, Biological Systems, Vehicular Networking and Connected Vehicles, Aerospace Systems, Automation, Manufacturing, Smart Grids, Nonlinear Systems, Power Systems, Robotics, Social Systems, Economic Systems and other. Of particular value to both the contributors and the readership are the short publication timeframe and the world-wide distribution and exposure which enable both a wide and rapid dissemination of research output.

The series covers the theory, applications, and perspectives on the state of the art and future developments relevant to systems and networks, decision making, control, complex processes and related areas, as embedded in the fields of interdisciplinary and applied sciences, engineering, computer science, physics, economics, social, and life sciences, as well as the paradigms and methodologies behind them.

Indexed by SCOPUS, EI Compendex, INSPEC, WTI Frankfurt eG, zbMATH, SCImago.

All books published in the series are submitted for consideration in Web of Science.

For proposals from Asia please contact Aninda Bose (aninda.bose@springer.com).

Akinori Ito
Editor

Proceedings of the 2025 10th International Conference on Multimedia Systems and Signal Processing

ICMSSP 2025, Fukui, Japan, May 9–11

Springer

Editor
Akinori Ito
Tohoku University
Miyagi, Japan

ISSN 2367-3370 ISSN 2367-3389 (electronic)
Lecture Notes in Networks and Systems
ISBN 978-3-032-05993-2 ISBN 978-3-032-05994-9 (eBook)
https://doi.org/10.1007/978-3-032-05994-9

This Springer imprint is published by the registered company Springer Nature Switzerland AG
The registered company address is: Gewerbestrasse 11, 6330 Cham, Switzerland

Preface

2025 10th International Conference on Multimedia Systems and Signal Processing (ICMSSP 2025) was held successfully in Fukui, Japan during May 9–11, 2025 with co-sponsored by University of Fukui, Japan and supported by Okayama University, Japan; Shonan Institute of Technology, Japan and Yamaguchi University, Japan.

The ICMSSP 2025 proceedings span four topic tracks: "Image model and calculation", "Image analysis and multimedia application", "Multimodal signal recognition and classification", and "Data-oriented information system design and innovation management". Each paper underwent a rigorous peer-review process conducted by expert referees to ensure the highest professional and scientific standards expected of the ICMSSP Technical Committees.

On behalf of the organizing committees, we would like to extend our deepest gratitude to all participants, including our keynote speakers, session chairs, authors, and delegates. We would also like to thank the reviewers for their insightful remarks and suggestions, which have enabled us to ensure the high quality of the publication. Finally, we thank Springer for their production of the elegant proceedings.

We believe that the scientific prototypes and technical content presented at ICMSSP 2025 will contribute to the continued success of Multimedia Systems and Signal Processing in industrial, entertainment, social, and everyday applications.

Akinori Ito
Conference Technical Program Chair

ICMSSP 2025 Conference Committees

Conference Advisory Committees

Wen-Chung Kao (Fellow, IEEE)	National Taiwan Normal University, Taiwan
Wen-Huang Cheng (Fellow, IEEE)	National Taiwan University, Taiwan
Hamadou Saliah-Hassane	TÉLUQ University, Canada
B. (Bedir) Tekinerdogan	Information Technology group at Wageningen University, The Netherlands

General Chairs

Nobuo Funabiki	Okayama University, Japan
Chih-Peng Fan	National Chung Hsing University, Taiwan

Local Arrangement Chair

Tomoya Kawakami	University of Fukui, Japan

Technical Program Chairs

Akinori Ito	Tohoku University, Japan
Shingo Yamaguchi	Yamaguchi University, Japan
Kiyoshi Ueda	Nihon University, Japan
Wilson Q. Wang	Lakehead University, Canada
Shinji Sugawara	Chiba Institute of Technology, Japan

Technical Program Co-chairs

Cheng Siong Chin	Newcastle University, Singapore
Masaru Fukushi	Yamaguchi University, Japan

Publication Chairs

Kazuhiro Kikuma	Nihon University, Japan
Kazuyuki Kojima	Shonan Institute of Technology, Japan

International Technical Committees

Yi Wang	Dongguan University of Technology, China
Li-Wen Chang	National Chung Hsing University, Taiwan
Jan-Ray Liao	National Chung Hsing University, Taiwan
Yu-Cheng Fan	National Taipei University of Technology, Taiwan
Li-Wei Kang	National Taiwan Normal University, Taiwan
Ritthiwut Puwaphat	Princess of Naradhiwas University, Thailand
Thai Hung Le	Vietnam National University, Vietnam
Nantapong Keandoungchun	King Mongkut's University of Technology Thonburi, Thailand
Naragain Phumchusri	Chulalongkorn University, Thailand
Ossama Embarak	Higher Colleges of Technology, UAE
John Blake	University of Aizu, Japan
Iara Margolis Ribeiro	University of Minho, Portugal
Yu-Mei Wang	University of Alabama at Birmingham, USA
Richard C. Larson	Massachusetts Institute of Technology, USA
Raymond Li	University of Canberra, Australia
Hacer Varol	Stephen F. Austin State University, USA
Yi-Fang Lee	National Taiwan Normal University, Taiwan
Loc Nguyen	Sunflower Soft Company, Vietnam
Nursabillilah Binti Mohd Ali	Universiti Teknikal Malaysia Melaka, Malaysia
P. K. Paul	Raiganj University, India
Emre Oner Tartan	Baskent University- Vocational School of Technical Sciences, Ankara, Turkey
Thaweesak Yingthawornsuk	King Mongkut's University of Technology Thonburi, Thailand
Gurendra Nath Bhardwaj	NIIT University, India
Radhakrishna Bhat	Manipal Institute of Technology, India
P. Geetha	Mohan Babu University, India
Nakhat Nasreen	Aligarh Muslim University, India
Shalini Garg	GGSIP University, India
Maritza Arones	Universidad Nacional "San Luis Gonzaga", Peru
Ajay Anil Gurjar	Sipna College of Engineering and Technology, India
Amitava Chatterjee	Jadavpur University, India

Contents

Multimodal Signal Recognition and Classification

Data-oriented Information System Design and Innovation Management

Image Model and Calculation

Fusion of Region-Constraint Attention and Convolution Neural Network for Blind Image Denoising

Jan-Ray Liao[1], Yen-Cheng Chang[1], and Li-Wen Chang[2](✉)

[1] Department of Electrical Engineering, National Chung Hsing University, 145 Xingda Road, South District, Taichung 402202, Taiwan
jrliao@mail.nchu.edu.tw

[2] General Education Center, Chaoyang University of Technology, 168 Jifeng East Road, Wufeng District, Taichung 413310, Taiwan
lwc@cyut.edu.tw

Abstract. Convolution layer and attention mechanism are two of the most important foundations for the success in image denoising through neural networks. Success of attention mechanism depends on the similarity score calculated from key and query generated from features of the image. As noise level increases, effectiveness of attention score drops because similarity matching becomes difficult. To overcome this performance reduction, traditional patch-based approaches performs matching in a small neighborhood to reduce noise interference. However, neural networks calculate attention score globally. This means that effect of noise interference is more severe in attention mechanism than traditional patch-based approaches. Therefore, in this paper, we propose to add a region constraint mask in attention calculation. With this newly added region constraint, attention score is calculated from a limited neighborhood around the considered pixel instead of from the entire image, thereby, reducing interference at high noise level. In addition, a convolutional path is added to complement the new region-constraint attentive path and these two parallel paths are fused together by another neural network to generate the final denoising result. With these two strategies, experimental results show that our proposed new network architecture performs better than state-of-the-art blind denoisers both quantitatively and qualitatively.

Keywords: denoising · region-constraint attention · convolution neural network

1 Introduction

Image denoising is a critical step in many image processing applications[1]. Before the rise of convolution neural networks (CNN) based methods [2], exploring non-local self similarity (NSS) in an image achieved the best performance in

A. Ito (Ed.): ICMSSP 2025, LNNS 1637, pp. 3–10, 2025.
https://doi.org/10.1007/978-3-032-05994-9_1

denoising [3,4]. The representative approaches in this category were non-local means (NLM) [3] and block-matching and 3D filtering (BM3D) [4]. The reason for their successes is by collecting structurally similar patches in an image and applying strategies such as weighted filtering or transform thresholding among these patch stacks. With the landmark development of DnCNN [2], denoisers based on CNN show significant performance advantage over these traditional patch-based approaches. The shortcoming of earlier CNN denoisers [5] is that they are non-blind. This means noise characteristics such as standard deviation needs to be known as a prior information. This requirement makes these earlier deep learning approaches impractical in real situations because it is almost impossible to know noise characteristics beforehand. Therefore, focus in denoising researches gradually turns to blind denoising without prior noise knowledge.

The basic components in a denoising neural network are convolution layer [2] and attention mechanism [6,7]. While convolution layer is used to extract useful features from images and reject noises in the process, attention mechanisms explore structural similarity in an image with motivations similar to traditional NSS methods. However, given the wide varieties of noise properties, using a single-path network by cascading these basic components is difficult to achieve satisfactory performance for blind denoising. Therefore, employing multiple strategies to handle varying noise characteristics is obviously preferable, i.e., a network should have multiple paths to perform different tasks. The first and the most obvious approach is to add noise estimation in a second path and use this additional information to assist non-blind denoising [8,9]. The second approach explores various convolution strategies [10,11]. The third approach utilizes attention mechanism [7,12], and transformer techniques [13–15].

Although attention mechanism shows great promises in image denoising as its counterpart in traditional NSS-based approaches, success of attention depends heavily on the score calculated from similarity between key and query extracted from the image. As noise level increases, the effectiveness of attention score reduces because similarity matching becomes more difficult. Although traditional patch-based approaches also face the same difficulty at high noise level, matching is done in a small neighborhood in traditional approaches and the effect of noise interference is reduced. On the other hand, attention score is calculated globally from the entire image in deep learning. This means that effect of noise interference is further increased in attention mechanism for CNN-based approaches. Given this issue, in this paper, we propose a fusion of two strategies to overcome the problem. The first is adding a convolution-based path in parallel to an attention-based path to reduce the effect of high noise interference. The rationale is that convolution-based approach is more similar to traditional filtering and it is more effective in high noise interference. Therefore, a convolutional path should be complementary to an attentive path and the results from the two parallel paths are fused together by another neural network to generate the final denoising result. The second is adding a region constraint mask in attention calculation. With region constraint, attention score is no longer calculated from an entire image but is obtained from a limited neighborhood, thereby, reducing interference from high noise level. With these two new strategies, experimental results show that the proposed new network architecture performs better than state-of-the-art blind denoisers quantitatively. Image quality of the new method is also better with less distortion than the other methods.

This paper is organized as follows: In the second section, we introduce the proposed network architecture. The third section presents experimental results and the fourth section concludes the paper.

2 Proposed Method

Image denoising problem can be described as an estimation from a noisy input: $\mathbf{Y} = \mathbf{X} + \mathbf{N}$ with $\mathbf{X}$ being the clean image and $\mathbf{N}$ being the additive noise. A denoising algorithm finds a mapping $\mathcal{F}$ estimates $\hat{\mathbf{X}}$ from $\mathbf{Y}$: $\hat{\mathbf{X}} = \mathcal{F}(\mathbf{Y})$ with the goal of minimizing differences between $\hat{\mathbf{X}}$ and $\mathbf{X}$. Therefore, we use mean squared error (MSE) loss as the optimization loss function without resorting to any other complicated loss functions. The loss function is $\mathcal{L} = \|\hat{\mathbf{X}} - \mathbf{X}\|^2$. In the following, we describe the network architecture that performs the denoising work in this paper:

2.1 Overall Architecture

The overall architecture of the network is shown in Fig. 1(a). As described in the introduction, the noisy input goes through a seven-layer convolution network shown in Fig. 1(b) and an attention network to be described later to generate $\hat{\mathbf{X}}_C$ and $\hat{\mathbf{X}}_A$, respectively. $\hat{\mathbf{X}}_C$, is also used as a complementary weight signal to generate two additional signals: $\hat{\mathbf{X}}_C \odot \hat{\mathbf{X}}_A$ and $\mathbf{Y} \odot (1 - \hat{\mathbf{X}}_C)$ where $\odot$ is the pixelwise multiplication between the two signals. The rationale for adding these two multiplications comes from the optimal denoising Wiener filter: $\frac{P_s(f)}{P_s(f)+P_n(f)}$ where $P_s(f)$ is the power spectrum of the undistorted signal and $P_n(f)$ is the power spectrum of the noise. Based on findings from previous work [8], $\hat{\mathbf{X}}_C \odot \hat{\mathbf{X}}_A$ and $\mathbf{Y} \odot (1 - \hat{\mathbf{X}}_C)$ are used as hints for signal and noise powers, thereby, incorporating the concept of Wiener filter into the network. The concatenation of the following signals: $\mathbf{G}_0 = [\hat{\mathbf{X}}_A, \hat{\mathbf{X}}_A \odot \hat{\mathbf{X}}_C, \hat{\mathbf{X}}_C, \mathbf{Y} \odot (1 - \hat{\mathbf{X}}_C), \mathbf{Y}])$ then goes through a three-layer convolution network to generate the final output.

2.2 Attention Network

Figure 1(c) shows the diagram of the attention network. It is inspired by residual non-local attention network (RNAN) [6] and made of a feature extraction convolution layer, an attentive denoising block, followed by five consecutive basic denoising blocks, another attentive denoising block, and a final reconstruction convolution layer.

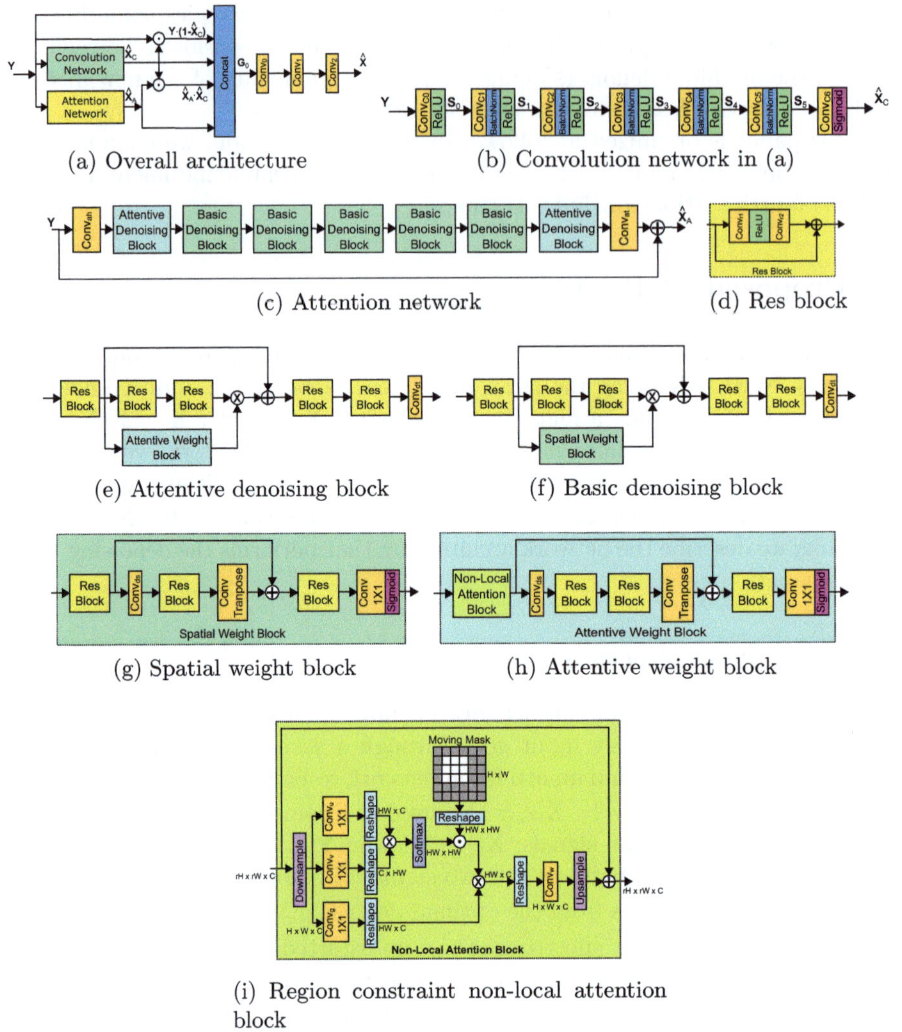

(a) Overall architecture

(b) Convolution network in (a)

(c) Attention network

(d) Res block

(e) Attentive denoising block

(f) Basic denoising block

(g) Spatial weight block

(h) Attentive weight block

(i) Region constraint non-local attention block

Fig. 1. Architecture of the proposed denoising network and diagrams of all the blocks.

The cores of the network are the attentive and basic denoising blocks as shown in Fig. 1(e) and (f). Both are made of canonical residual block in Fig. 1(d). The difference between attentive and basic denoising blocks are the weight blocks used for generating a spatial weight map to be multiplied with residual features in order to either strengthen or weaken these features for denoising. The attentive and spatial weight blocks are shown in Fig. 1(g) and (h), respectively.

2.3 Region Constraint Non-local Attention Block

Figure 1(i) shows the diagram of the region constraint non-local attention block. In a typical attention block, the input passes through convolution layers Conv_u, Conv_v, and Conv_g to generate query, key, and value matrices. The product of query and key is passed through softmax function to produce an attention score matrix. The score is multiplied with the value to produce the attention weighted result, then, passes through convolution layer Conv_w, and, finally, adds with the input to generate the output.

The differences of our region constraint non-local attention block and the typical attention block are twofold. First, downsample and upsample blocks are added at the start and end of the attention block. The reason for adding these two blocks is because the complexity of an attention block is proportional to $(HW)^2$ where H and W are the height and width of the input. When image size exceeds a certain threshold, memory requirements may exceed the limit of hardware. Therefore, an adjustable factor r is introduced in the downsample and upsample blocks so that memory requirements are within hardware limitation. Second, a moving mask is applied to the attention score matrix after the softmax operation. In a typical attention block, the attention score is applied to all the features in the value matrix globally. In the newly introduced mask matrix, each row corresponds to a region constraint moving mask centered at the corresponding location of the value matrix. These two-dimensional moving masks are reshaped into one-dimensional mask vectors. The end result is that the attention score is now applied to a region surrounding the location of each input feature. Therefore, a region-constraint non-local attention is generated which is more similar to traditional patch-based matching approaches instead of a typical global attention approach.

3 Experimental Results

3.1 Experimental Setup

We used images in Berkeley Segmentation Data Set and Benchmarks 500 (BSDS500) [16] for training. Image patches of size 180×180 are randomly extracted from the training images. For each patch, zero-mean Gaussian noise is added with standard deviation randomly selected in the range of 0 to 75. We train the network for 300 epochs using Adam optimizer. The initial learning rate is 0.001 and it is halved every 40 epochs. The hardware platform used is Intel i7-8700 3.2GHz CPU with 12GB RAM and NVIDIA GeForce RTX 3070 with 8GB RAM. The software environment is Microsoft Windows 10, Python 3.6.13, PyTorch 1.9.0, and CUDA 11.1.

We compare our algorithm with the state-of-the-art algorithms: DnCNN [2], FFDNet [5], BUIFD [8], and RDN [17]. Although FFDNet is not a blind denoising algorithm, it is included because of its significance in denoising researches.

Images in Set12 and Kodak24 datasets are used in testing. Peak signal-to-noise ratio (PSNR), structural similarity index measure (SSIM) [18], and LPIPS [19] are used for quantitative comparison.

3.2 Denoising Performance

σ \ PSNR↑	DnCNN	FFDNet	BUIFD	RDN	Proposed
15	32.49	**32.68**	32.38	32.14	32.41
35	28.09	**28.63**	28.32	28.19	28.48
55	25.25	25.97	25.70	25.89	**26.05**
75	23.24	23.47	23.61	23.78	**24.00**
σ \ SSIM↑	**DnCNN**	**FFDNet**	**BUIFD**	**RDN**	**Proposed**
15	0.895	**0.902**	0.896	0.891	0.897
35	0.796	**0.823**	0.812	0.808	0.821
55	0.705	0.753	0.730	0.748	**0.756**
75	0.641	0.680	0.665	0.692	**0.703**
σ \ LPIPS↓	**DnCNN**	**FFDNet**	**BUIFD**	**RDN**	**Proposed**
15	**0.085**	0.102	0.103	0.086	**0.085**
35	0.159	0.185	**0.151**	0.168	0.159
55	0.253	0.272	0.218	0.238	**0.211**
75	0.317	0.367	0.286	0.297	**0.265**

(a) Set12

σ \ PSNR↑	DnCNN	FFDNet	BUIFD	RDN	Proposed
15	32.35	**32.63**	32.42	31.68	32.35
35	27.75	**28.47**	28.24	28.00	28.40
55	24.98	25.87	25.63	25.77	**26.03**
75	23.07	23.57	23.59	23.79	**24.07**
σ \ SSIM↑	**DnCNN**	**FFDNet**	**BUIFD**	**RDN**	**Proposed**
15	0.876	**0.888**	0.883	0.867	0.882
35	0.736	**0.783**	0.772	0.755	**0.783**
55	0.635	0.699	0.677	0.683	**0.712**
75	0.572	0.630	0.608	0.639	**0.660**
σ \ LPIPS↓	**DnCNN**	**FFDNet**	**BUIFD**	**RDN**	**Proposed**
15	0.118	0.124	0.125	0.128	**0.103**
35	0.258	0.249	0.233	0.267	**0.226**
55	0.373	0.380	0.334	0.371	**0.308**
75	0.452	0.491	0.411	0.431	**0.377**

(b) Kodak24

Fig. 2. PSNRS (dB), SSIM, and LPIPS of denoising algorithms under different σ for datasets Set12 and Kodak24. Bold indicates the top value. Underline indicates the second value from the top.

Figure 2 shows PSNR, SSIM, and LPIPS for Set12 and Kodak24 datasets. We can see that our proposed method outperforms all the other algorithms and gives the best results at high noise levels. At low noise levels, our proposed method is still among the top two performers except PSNR at $\sigma = 15$. Figure 3 shows perceptual comparison for image 4 of Set12 dataset and image 5 of Kodak24 dataset. From the figure, we can see that the proposed algorithm removes noise well and preserves details with less distortion as compared to the other algorithms. This is especially noticeable when compared to FFDNet and RDN. As we can see from the figure, the proposed method has far less blurring than FFDNet and RDN.

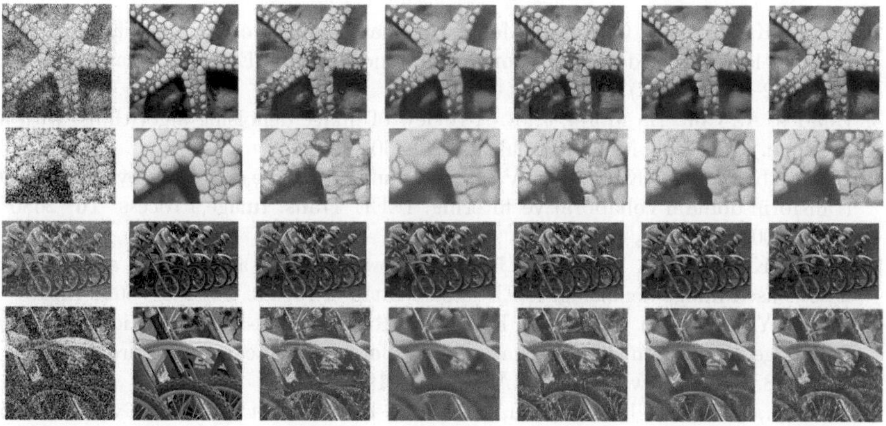

Fig. 3. Comparison of the denoised outputs for image 4 in Set12 dataset (1st and 2nd rows) and image 5 in Kodak24 dataset (3rd and 4th rows) at $\sigma = 75$. From left to right are: noisy image, original image, DnCNN, FFDNet, BUIFD, RDN, and proposed method. The 2nd and 4th rows show the enlarged region marked in the red rectangle of the original image.

4 Conclusion

Following the success of patch matching in traditional image denoising algorithms, attention mechanism is its counterpart in neural networks. Similar to patch matching, success of attention mechanism depends on the similarity score calculated from key and query generated from the image. As noise level increases, similarity matching becomes difficult and effectiveness of attention score drops. To overcome this issue, traditional patch matching limits searching to a small neighborhood. Since attention score is calculated globally in neural networks, effect of noise interference is more severe in attention mechanism. Therefore, in this paper, we propose to add a region constraint mask in attention calculation to overcome this issue. With this new mechanism, attention score is calculated from a limited neighborhood instead of from the entire image, thereby, reducing interference at high noise level. In addition, we add a convolutional path to complement the attentive path and the two parallel paths are fused together by another neural network to generate the final denoising result. With these two strategies, experimental results show that our proposed new network architecture performs better than state-of-the-art blind denoisers both quantitatively and qualitatively.

References

1. Li, Y., Zhang, Y., Timofte, R., Gool, L.V., Tu, Z., et al.: Ntire 2023 challenge on image denoising: Methods and results. In: IEEE/CVF Conference on Computer Vision and Pattern Recognition Workshops (CVPRW), pp. 1905–1921. IEEE, Vancouver, BC, Canada (2023)

2. Zhang, K., Zuo, W., Chen, Y., Meng, D., Zhang, L.: Beyond a gaussian denoiser: residual learning of deep CNN for image denoising. IEEE Trans. Image Process. **26**, 3142–3155 (2017)
3. Buades, A., Coll, B., Morel, J.M.: A review of image denoising algorithms, with a new one. Multiscale Model. Simul. **4**, 490–530 (2005)
4. Dabov, K., Foi, A., Katkovnik, V., Egiazarian, K.: Image denoising by sparse 3-D transform-domain collaborative filtering. IEEE Trans. Image Process. **16**, 2080–2095 (2007)
5. Zhang, K., Zuo, W., Zhang, L.: FFDNet: toward a fast and flexible solution for CNN based image denoising. IEEE Trans. Image Process. **27**, 4608–4622 (2018)
6. Zhang, Y., Li, K., Li, K., Zhong, B., Fu, Y.: Residual non-local attention networks for image restoration. In: International Conference on Learning Representations, pp. 1–18. ICLR, New Orleans, LA, USA (2019)
7. Mou, C., Zhang, J., Fan, X., Liu, H., Wang, R.: Cola-net: collaborative attention network for image restoration. IEEE Trans. Multimedia **24**, 1366–1377 (2022)
8. El Helou, M., Susstrunk, S.: Blind universal bayesian image denoising with gaussian noise level learning. IEEE Trans. Image Process. **29**, 4885–4897 (2020)
9. Pan, Y., Ren, C., Wu, X., Huang, J., He, X.: Real image denoising via guided residual estimation and noise correction. IEEE Trans. Circuits Syst. Video Technol. **33**, 1994–2000 (2023)
10. Zamir, S.W., et al.: Learning enriched features for fast image restoration and enhancement. IEEE Trans. Pattern Anal. Mach. Intell. **45**, 1934–1948 (2023)
11. Wu, W., Liu, S., Xia, Y., Zhan, Y.: Dual residual attention network for image denoising. Pattern Recogn. **149**, 110291 (2024)
12. Mei, Y., et al.: Pyramid attention network for image restoration. Int. J. Comput. Vision **131**, 3207–3225 (2023)
13. Zamir, S.W., Arora, A., Khan, S.H., Hayat, M., Khan, F.S., Yang, M.H.: Restormer: efficient transformer for high-resolution image restoration. In: IEEE/CVF Conference on Computer Vision and Pattern Recognition, pp. 5728–5739. IEEE, New Orleans, LA, USA (2022)
14. Xiao, J., Fu, X., Zhou, M., Liu, H., Zha, Z.J.: Random shuffle transformer for image restoration. In: 40th International Conference on Machine Learning, pp. 38039–38058. ICML, Honolulu, Hawaii (2023)
15. Zhang, J., Zhang, Y., Gu, J., Zhang, Y., Kong, L., Yuan, X.: Accurate image restoration with attention retractable transformer. In: 11th International Conference on Learning Representations (ICLR), pp. 1–13. ICLR, Kigali, Rwanda (2023)
16. Arbelaez, P., Maire, M., Fowlkes, C., Malik, J.: Contour detection and hierarchical image segmentation. IEEE Trans. Pattern Anal. Mach. Intell. **33**, 898–916 (2011)
17. Zhang, Y., Tian, Y., Kong, Y., Zhong, B., Fu, Y.: Residual dense network for image restoration. IEEE Trans. Pattern Anal. Mach. Intell. **43**, 2480–2495 (2021)
18. Wang, Z., Bovik, A.C., Sheikh, H.R., Simoncelli, E.P.: Image quality assessment: from error visibility to structural similarity. IEEE Trans. Image Process. **13**, 600–612 (2012)
19. Zhang, R., Isola, P., Efros, A.A., Shechtman, E., Wang, O.: The unreasonable effectiveness of deep features as a perceptual metric. In: IEEE/CVF Conference on Computer Vision and Pattern Recognition, pp. 586–595. IEEE, Salt Lake City, UT, USA (2018)

Learning Contour-Guided 3D Face Reconstruction with Occlusions

Dapeng Zhao and Xiaoran Yan(✉)

Zhejiang Lab, Hangzhou 311121, China
yanxr@zhejianglab.com

Abstract. 3D face reconstruction is a versatile technology applied in various contexts. Deep learning methods, in particular, prove highly valuable in this domain due to their ability to generate high-quality 3D models. Nevertheless, most of these techniques require unobstructed and clear facial images as input. In response, we've developed a 3D face reconstruction system capable of performing effectively even when faced with obstructed views. Taking inspiration from generative face image inpainting, we've introduced a comprehensive method for crafting facial representations guided by their outlines. Our network comprises two components: one dedicated to image enhancement by removing obstructions and restoring missing facial features, while the other utilizes weak supervision to craft exceptionally detailed 3D models. Through extensive experimentation on standard 3D face reconstruction tasks, we've demonstrated the superiority of our method compared to existing ones that often fall short. Our results, based on experiments conducted with LFW databases, affirm the effectiveness of our approach.

Keywords: 3D Face Reconstruction · Face Recognition · Occluded Scenes

1 Introduction

Face reconstruction systems find extensive applications in specific scenarios, including face recognition [1, 2] and digital entertainment [3]. Nonetheless, the majority of these systems require the input face photos to be unobstructed and depict the front of the face. In contrast to recent research efforts focused on enhancing facial resolution and fidelity [4], 3D face reconstruction in occluded environments remains underexplored. Consequently, addressing the challenge of reconstructing faces in obstructed scenes is a pressing task.

Face images can be occluded by anything, such as palms, glasses, or leaves. Without any prior or weak supervision, it has been shown that we cannot simply infer the occluded face image. However, face image synthesis has achieved tremendous success in recent years due to the rapid development of Generative Adversarial Networks (GANs). State-of-the-art GAN techniques, such as Stylegan2 [5], can generate high-fidelity virtual face images that are sometimes even difficult to distinguish from real ones.

In this work, we investigated the identification of occluded areas and the removal of occlusions to synthesize a complete face from a human face. We used face parsing

A. Ito (Ed.): ICMSSP 2025, LNNS 1637, pp. 11–20, 2025.
https://doi.org/10.1007/978-3-032-05994-9_2

networks to identify the occluded areas and used a contour-based GAN to synthesize a complete 2D face. Due to the lack of 3D datasets, we utilized a weakly supervised deep learning framework to predict the coefficients of the 3D Morphable Model (3DMM). In the end, we obtained the final face model with complete facial features.

This Research Makes Three Contributions:

- We propose an algorithm that combines face analysis map and face contour map to generate face with full facial features.
- We have improved the loss function of our 3D reconstruction framework for occluded scenes. Our results are more accurate than other state-of-the-art methods.
- We propose a novel 3D face reconstruction method that can produce accurate models under occluded scenes.

2 Related Work

2.1 Generic Face Reconstruction

Deep learning-based methods [6] directly regress 3D Morphable Model (3DMM) coefficients from images. Some frameworks [7] have explored end-to-end convolutional neural network (CNN)architectures to regress 3DMM coefficients directly. However, most of these methods require that the input image is an unobstructed frontal face. In order to obtain paired 2D-3D data for supervised learning, these approaches typically require very deep CNNs, which can be difficult to train. Additionally, these approaches do not perform well when dealing with complex lighting, occlusion, and other in-the-wild conditions.

2.2 Face Parsing

Face parsing maps are commonly used as an essential intermediate representation for generative modeling [8]. They can be used as the conditional context for conditional face image synthesis. Image-to-image translation models can learn the mapping from parsing maps to realistic RGB images [9]. A practical face parsing solution should directly predict per-pixel semantic labels across the entire face image. To obtain accurate parsing maps, Wei *et al.* [10] developed a novel method for regulating receptive fields with superior regulation ability in parsing networks. MaskGan [11] provided a labeled face parsing dataset. Zhao *et al.* [12] developed a model that explored how to combine the fully convolutional network and super-pixels data. To address the problem of limited access to global image information, some methods [13] have introduced the transformer component. Semantic layouts provide rough guidance of the location and appearance of objects, which further facilitates training.

2.3 Face Image Synthesis

Due to the rapid development of convolutional neural networks, deep pixel-level face generation has shown extraordinary capabilities. Recently, several methods have been introduced that pro-vide additional information before face image synthesis. One of the first deep learning methods designed for image inpainting is context encoder [14], which uses an encoder-decoder framework. However, the network performs poorly on human faces. Following this work, Dolhansky *et al.* [15] demonstrated the importance of exemplar data for inpainting. However, this method only focuses on filling in missing eye regions of frontal faces, so it does not generalize well. In subsequent work, a pre-trained VGG network was used to improve the output of the context encoder by minimizing the feature differences between the input image and the inpainted image [16].

In the face synthesis task, the encoder converts an image with the concerned regions into a low-dimensional feature space. The decoder then builds the output image. However, due to the data bottleneck in the channel-wise fully connected layer, the restored regions of the output image often contain visual artifacts and appear blurry. Iizuka *et al.* [17] solved this problem by replacing the fully connected layer with a series of dilated convolutional layers. Yu *et al.* [18] proposed a two-step image synthesis method. In the first step, it estimates the missing area in a coarse manner. The refinement network then uses an attention mechanism to sharpen the result by looking for a collection of background patches with the highest similarity to the coarse estimate.

3 Proposed Approach

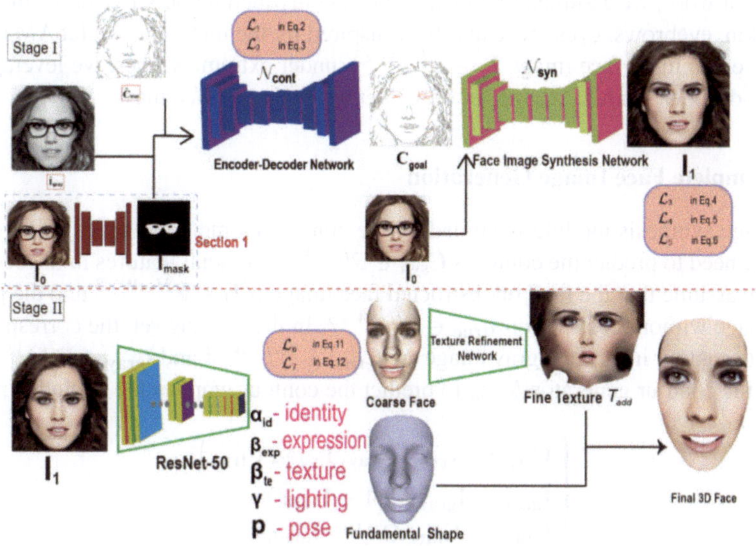

Fig. 1. Method overview. See related sections for details.

We propose a 3D face reconstruction method based on a single image (as shown in Fig. 1. The method consists of two stages. The first stage, face image synthesis under occluded scenes, synthesizes a complete face image from an occluded image. The second stage, a 3D face reconstruction network based on unobstructed frontal images, reconstructs a 3D face model from the synthesized complete face image. Our goal is to realize 3D face reconstruction under occluded scenes using this framework. Given a source image $\mathbf{I_{input}} \in \mathbb{R}^{H\times W\times 3}$ with obstructions on the face, we obtain the final face model.

3.1 How to Obtain the Face Mask

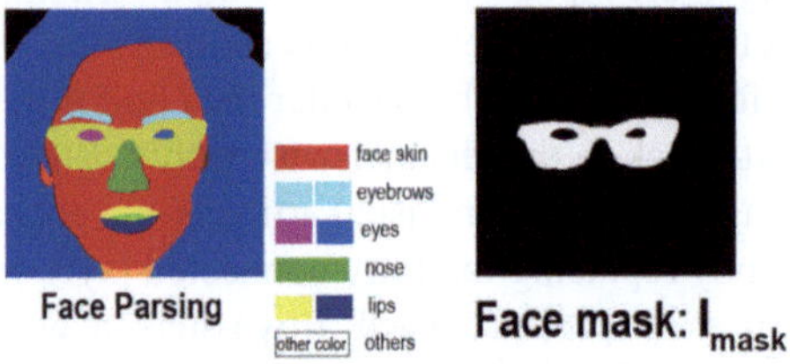

Fig. 2. The face mask we define, indicating the position of the occlusions.

As show in Sect. 1 of Fig. 1 and Fig. 2, we define face mask image $\mathbf{I_{mask}} \in \mathbb{R}^{H\times W\times 1}$ (1 for the occluded occlusions, 0 for background) indicating the position of the occlusions. Our face mask $\mathbf{I_{mask}}$ and traditional face parsing map are closely related. We need to separate the facial features, and we can reveal the area of the occlusions. Please notice that, in our work, we assumed that facial features only include only five parts, including facial skin, eyebrows, eyes, nose and lips. Inspired by the method of CelebAMask-HQ [11], given an input face image $\mathbf{I_0} \in \mathbb{R}^{H\times W\times 3}$ under extreme scenes, we leveraged an encoder-decoder module Ω_m based on U-Net to obtain the face mask.

3.2 Complete Face Image Generation

Our image synthesis module is guided by the contour (some approaches called *edge*). First, we need to predict the contours $\mathbf{C_{syn}} \in \mathbb{R}^{H\times W\times 1}$ of facial features in the occluded area. We assume that the final unobstructed face image is $\mathbf{I_1} \in \mathbb{R}^{H\times W\times 3}$ and the ground truth image without obstruction $\mathbf{I_{true}} \in \mathbb{R}^{H\times W\times 3}$. In the training set, the corresponding complete contour image and gray image are $\mathbf{C_{true}} \in \mathbb{R}^{H\times W\times 1}$ and $\mathbf{I_{gray}} \in \mathbb{R}^{H\times W\times 1}$. We trained the contour generator $\mathcal{N}_{cont}$ to predict the contour map for the occluded region:

$$\begin{cases} \mathbf{C_{syn}} = \mathcal{N}_{cont}\left(\hat{\mathbf{I}}_{\mathbf{gray}}, \mathbf{I_{mask}}, \hat{\mathbf{C}}_{\mathbf{true}}\right) \\ \hat{\mathbf{I}}_{\mathbf{gray}} = \mathbf{I_{gray}} \odot (1 - \mathbf{I_{mask}}) \\ \hat{\mathbf{C}}_{\mathbf{ture}} = \mathbf{C_{ture}} \odot (1 - \mathbf{I_{mask}}) \end{cases}$$

where $\hat{\mathbf{I}}_{\mathbf{gray}}$ denotes masked grayscale image, $\hat{\mathbf{C}}_{\mathbf{ture}}$ denotes masked contour image and $\odot$ denotes the Hadamard product.

We trained the discriminator of the module $\mathcal{N}_{cont}$ to predict which of $\mathbf{C_{syn}}$ and $\mathbf{C_{ture}}$ is true contour map and which is false contour map. The adversarial loss is defined as

$$\mathcal{L}_1 = \mathbb{E}_{(\mathbf{C_{ture}},\mathbf{I_{gray}})}\left[\log D_1\left(\mathbf{C_{true}}, \mathbf{I_{gray}}\right)\right] + \mathbb{E}_{(\mathbf{I_{gray}})} \log\left[1 - D_1\left(\mathbf{C_{syn}}, \mathbf{I_{gray}}\right)\right]$$

In addition, we compare the feature activation maps of the discriminator. We set the face feature matching loss as

$$\mathcal{L}_2 = \mathbb{E}\left[\sum_{i=1}^{K} \frac{1}{N_i} \left\| D_1^{(i)}(\mathbf{C_{true}}) - D_1^{(i)}(\mathbf{C_{syn}}) \right\| \right]$$

where N_i is the number of elements in the ith activation layer, K is the final convolution layer of the discriminator and $D_1^{(i)}$ is the activation in the ith layer of the discriminator.

After obtaining the complete contour map, we design $\mathcal{N}_{syn}$ to generate the complete face image $\mathbf{I_1}$. The complete contour map $\mathbf{C_{goal}}$ is formed by adding $\mathbf{C_{syn}}$ and $\mathbf{C_{true}}$, which follows $\mathbf{C_{goal}} = \mathbf{C_{true}} \odot (1 - \mathbf{I_{mask}}) + \mathbf{C_{syn}} \odot \mathbf{I_{mask}}$. In the map $\mathbf{C_{goal}}$, we can see the contours of all facial features, especially the occluded areas. In addition, we set $\hat{\mathbf{I}}_{\mathbf{true}} \in \mathbb{R}^{H \times W \times 3}$ to be an incomplete face picture, which follows $\hat{\mathbf{I}}_{\mathbf{true}} = \mathbf{I_{true}} \odot (1 - \mathbf{I_{mask}})$. So, we utilize $\mathcal{N}_{syn}$ to get the final complete face image $\mathbf{I_1}$, with occluded regions recovered, which follows $\mathbf{I_1} = \mathcal{N}_{syn}(\hat{\mathbf{I}}_{\mathbf{true}}, \mathbf{C_{goal}})$.

We trained the module $\mathcal{N}_{syn}$ to predict the final complete face image $\mathbf{I_1}$ over a joint loss. The adversarial loss is defined as

$$\mathcal{L}_3 = \mathbb{E}_{(\mathbf{I_{true}}, \mathbf{C_{fina}})}\left[\log D_2\left(\mathbf{I_{true}}, \mathbf{C_{goal}}\right)\right] + \mathbb{E}_{(\mathbf{C_{fina}})} \log\left[1 - D_2\left(\mathbf{I_1}, \mathbf{C_{goal}}\right)\right]$$

The per-pixel loss is defined as follows

$$\mathcal{L}_4 = \frac{1}{S_m} \|\mathbf{I_1} - \mathbf{I_{true}}\|_1$$

which S_m denotes the size of the face mask $\mathbf{I_{mask}}$ and $\|\cdot\|_1$ denotes the L_1 norm. Notice that we use the mask size S_m as the denominator to adjust the penalty.

The style loss [19] computes the style distance between two face images as follows

$$\mathcal{L}_5 = \sum_{n} \frac{1}{Q_n \times Q_n} \left\| \frac{G_n(\mathbf{I_1} \odot (1 - \mathbf{I_{mask}})) - G_n(\hat{\mathbf{I}}_{\mathbf{true}})}{Q_n \times H_n \times W_n} \right\|_1$$

where $G_n(x) = \varphi_n(x)^T \varphi_n(x)$ denotes the Gram Matrix corresponding to $\varphi_n(x)$, $\varphi_n(\cdot)$ denotes the Q_n feature maps with the size $H_n \times W_n$ of the n-th layer. In summary, the contour generator network $\mathcal{N}_{cont}$ was trained with an objective comprised of an adversarial loss and feature-matching loss

$$\min_{G_1} \max_{D_1} \mathcal{L}_{G_1} = \min_{G_1}\left(\lambda_1 \max_{D_1} \mathcal{L}_1 + \lambda_2 \mathcal{L}_2\right)$$

The total loss function of $\mathcal{N}_{syn}$ follows

$$\min_{G_2} \max_{D_2} \mathcal{L}_{G_2} = \lambda_3 \max_{D_2} \mathcal{L}_3 + \lambda_4 \mathcal{L}_4 + \lambda_5 \mathcal{L}_5$$

here, we set $\lambda_1 = 1$, $\lambda_2 = 11.5$, $\lambda_3 = 0.1$, $\lambda_4 = 1$ and $\lambda_5 = 250$ respectively.

3.3 3D Model Reconstruction

Since the face is a very standardized figure, we use a human face template to construct the basic shape (here, we adopt 3DMM) [20]. Commonly, we describe single people's face with PCA [1], where shape and texture are separated:

$$\begin{cases} \mathbf{S} = \overline{\mathbf{S}} + \mathbf{A}_{\mathbf{id}}\boldsymbol{\alpha}_{\mathbf{id}} + \mathbf{B}_{\mathbf{exp}}\boldsymbol{\beta}_{\mathbf{exp}} \\ \mathbf{T} = \overline{\mathbf{T}} + \mathbf{B}_{\mathbf{te}}\boldsymbol{\beta}_{\mathbf{te}} \end{cases}$$

where $\overline{\mathbf{S}}$ and $\overline{\mathbf{T}}$ denote the average shape and texture, $\mathbf{A_{id}}$, $\mathbf{B_{exp}}$ and $\mathbf{B_{te}}$ denote the PCA basis of shape (identity and expression) and texture. $\boldsymbol{\alpha}_{\mathbf{id}} \in \mathbb{R}^{80}$, $\boldsymbol{\beta}_{\mathbf{exp}} \in \mathbb{R}^{64}$ and $\boldsymbol{\beta}_{\mathbf{te}} \in \mathbb{R}^{80}$ are the corresponding 3DMM coefficient vectors. After the 3D face is reconstructed, it can be projected onto the image plane with the perspective projection:

$$\mathbf{P_{2d}} = k * \mathbf{P_p} * \mathbf{R} * \mathbf{S_{mod}} + \mathbf{t_{2d}}$$

where $\mathbf{P_{2d}}$ denotes the projection function that turned the 3D model into 2D face positions, k denotes the scale factor, $\mathbf{P_p}$ denotes the projection matrix, $\mathbf{R} \in SO(3)$ denotes the rotation matrix and $\mathbf{t_{2d}} \in \mathbb{R}^3$ denotes the translation vector.

We approximated the scene illumination with Spherical Harmonics (SH) [21] parameterized by coefficient vector $\gamma \in \mathbb{R}^9$. Overall, the unknown vector parameters can be formulated $\mathbf{V_x} = (\boldsymbol{\alpha}_{\mathbf{id}}, \boldsymbol{\beta}_{\mathbf{exp}}, \boldsymbol{\beta}_{\mathbf{te}}, \boldsymbol{\gamma}, \mathbf{p}) \in \mathbb{R}^{239}$, where $\mathbf{p} \in \mathbb{R}^6 = \{\mathbf{pitch}, \mathbf{yaw}, \mathbf{roll}, k, \mathbf{t_{2D}}\}$ denotes face poses. Here, we used a revised ResNet-50 [22] network to predict $\mathbf{V_x}$. The corresponding loss function consists of two parts: pixel-wise loss and face feature loss.

Per-Pixel Loss. The pixel loss function minimizes the difference between the input image $\mathbf{I}_{\mathbf{fina}}^{(j)}$ and the output image $\mathbf{I}_{\mathbf{y}}^{(j)}$. The rendering layer renders back a rendered image $\mathbf{I}_{\mathbf{y}}^{(j)}$ to compare with the image $\mathbf{I}_{\mathbf{1}}^{(j)}$. We compute the per-pixel loss with:

$$\mathcal{L}_6 = \left\| \mathbf{I}_{\mathbf{1}}^{(j)} - \mathbf{I}_{\mathbf{y}}^{(j)} \right\|_2$$

where j denotes pixel index and $\|\cdot\|_2$ denotes the L_2 norm.

Face Features Loss We introduce a loss function at the face recognition level to reduce the difference between the 3D model of the face and the 2D image. The loss function computes the feature difference between the input image $\mathbf{I_1}$ and rendered image $\mathbf{I_y}$. We define the loss as a cosine distance:

$$\mathcal{L}_7 = 1 - \frac{<G(\mathbf{I_1}), G(\mathbf{I_y})>}{\|G(\mathbf{I_1})\| \cdot \|G(\mathbf{I_y}\|}$$

where $G(\cdot)$ denotes the feature extraction function by FaceNet [23], $< \cdot, \cdot >$ denotes the inner product.

In summary, we used the loss function $\mathcal{L}_{3D}$ to reconstruct the basic shape of the face. We set $\mathcal{L}_{3D} = \lambda_6\mathcal{L}_6 + \lambda_7\mathcal{L}_7$, where $\lambda_6 = 1.4$ and $\lambda_7 = 0.25$ respectively in all our experiments. We then used a coarse-to-fine graph convolutional network based on the frameworks of Lin *et al.* [4] for producing the fine texture T_{add}.

4 Experimental Results

4.1 Qualitative Comparison with Other Methods

Figure 3 shows our experimental results compared with the others [24]. The result shows that our method is far superior to other frameworks. Our 3D reconstruction method can handle occluded scenes, such as palms, chains on the forehead, and glasses. Other frameworks can not handle occlusions well; they are more focused on the generation of high-definition textures.

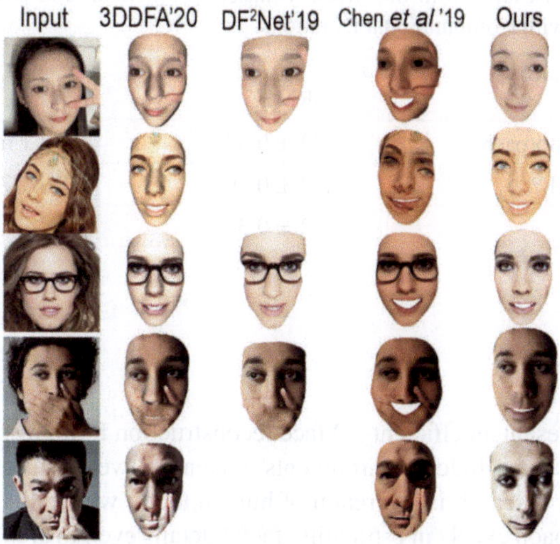

Fig. 3. Comparison of qualitative results. Baseline methods from left to right: 3DDFA, DF^{2}Net, Chen *et al.*, and our method. The blank area means that the corresponding method does not work.

4.2 Quantitative Comparison with Other

To thoroughly assess the advantages of our proposed approach and to highlight its robustness and flexibility with unconstrained facial imagery, we employed the MICC dataset [25] as the evaluation standard. Through meticulous experimental analysis, we juxtaposed our technique against a Structure-from-Motion (SFM) approach [2, 26–28] and a learning-driven method [6, 29]. The MICC dataset comprises 53 video sequences, each varying in resolution, capture conditions, and scaling factors. Data were gathered across controlled, semi-controlled, and entirely uncontrolled settings to guarantee the diversity and representativeness of the experimental scenarios. For each participant, the dataset supplies a ground truth model derived from structured light scanning, serving as the reference for gauging reconstruction precision. Reconstruction errors were computed based on the aforementioned techniques. During the experiments, we extracted the most frontal facial image from each participant's outdoor video sequence and utilized it as input to generate a3D facial model using our method. This procedure further substantiates the algorithm's efficacy and resilience. Table 1 reveals that our reconstruction error is comparable to that of Tran *et al.*'s approach [6] while outperforming the method proposed by Hernandez *et al.* [26]. This disparity can be attributed to the inherent constraints of Hernandez *et al.*'s SFM technique, which excessively relies on the continuity of image sequences and assumptions regarding camera motion. Although Tran *et al.*'s method is similarly hampered by a pronounced dependence on the distribution of training data and limited generalizability to unseen samples, its error rate remains on par with ours. By leveraging dependable prior knowledge of intermediate facial structures and SFS recovery techniques, our approach effectively achieves robust estimation on unconstrained images.

Table 1. Numerical Results from the MICC Datasets [25]: Mean and Standard Deviation of 3DRMSE, Coupled with Computational Durations.

Method	3DRMSE	run time
Hernandez *et al.*'s approach	1.92 ± 0.39	CPU 1 min 13 s
Tran *et al.*'s approach	1.53 ± 0.29	GPU 1.77 s
Ours	1.62 ± 0.37	CPU 1 min 40 s

5 Conclusions

In this study, we present an efficient 3D face reconstruction framework capable of functioning effectively in occluded environments. Given the wealth of domain knowledge and extensive prior research in the realm of human faces, we readily acquired relevant datasets. Our approach excels in estimating facial details even in areas where the face is obscured, such as by hands, forehead accessories, or glasses. Extensive experiments have conclusively demonstrated that our method significantly outperforms previous approaches in both accuracy and robustness.

Acknowledgments. Supported by National Key R&D Program of China (No. 2022YFF0712400).

References

1. Blanz, V., Vetter, T.: Face recognition based on fitting a 3D morphable model. IEEE Trans. Pattern Anal. Mach. Intell. **25**(9), 1063–1074 (2003)
2. Zhang, Y., et al.: Temporal transformer networks with self-supervision for action recognition. IEEE Internet Things J. **10**(14), 12999–13011 (2023)
3. Cao, C., Hou, Q., Zhou, K.: Displaced dynamic expression regression for real-time facial tracking and animation. ACM Trans. Graph. (TOG) **33**(4), 1–10 (2014)
4. Lin, J., Yuan, Y., Shao, T., Zhou, K.: Towards high-fidelity 3D face reconstruction from in-the-wild images using graph convolutional networks. arXiv preprint arXiv:2003.05653 (2020)

5. Karras, T., Laine, S., Aittala, M., Hellsten, J., Lehtinen, J., Aila, T.: Analyzing and improving the image quality of StyleGAN. In: Proceedings of the IEEE/CVF Conference on Computer Vision and Pattern Recognition, pp. 8110–8119 (2020)
6. Tran, A.T., Hassner, T., Masi, I., Medioni, G.: Regressing robust and discriminative 3D morphable models with a very deep neural network. In: Proceedings of the IEEE Conference on Computer Vision and Pattern Recognition, pp. 5163–5172 (2017)
7. Jourabloo, A., Liu, X.: Large-pose face alignment via CNN-based dense 3D model fitting. In: Proceedings of the IEEE Conference on Computer Vision and Pattern Recognition, pp. 4188–4196 (2016)
8. Johnson, J., Gupta, A., Fei-Fei, L.: Image generation from scene graphs. In: Proceedings of the IEEE Conference on Computer Vision and Pattern Recognition, pp. 1219–1228 (2018)
9. Isola, P., Zhu, J.-Y., Zhou, T., Efros, A.A.: Image-to-image translation with conditional adversarial networks. In: Proceedings of the IEEE Conference on Computer Vision and Pattern Recognition, pp. 1125–1134 (2017)
10. Wei, Z., Sun, Y., Wang, J., Lai, H., Liu, S.: Learning adaptive receptive fields for deep image parsing network. In: Proceedings of the IEEE Conference on Computer Vision and Pattern Recognition, pp. 2434–2442 (2017)
11. Lee, C.-H., Liu, Z., Wu, L., Luo, P.: MaskGAN: towards diverse and interactive facial image manipulation. In: Proceedings of the IEEE/CVF Conference on Computer Vision and Pattern Recognition, pp. 5549–5558 (2020)
12. Zhou, L., Liu, Z., He, X.: Face parsing via a fully-convolutional continuous CRF neural network. arXiv preprint arXiv:1708.03736 (2017)
13. Yin, Z., Yiu, V., Xiaolin, H., Tang, L.: End-to-end face parsing via interlinked convolutional neural networks. Cogn. Neurodyn. **15**(1), 169–179 (2021)
14. Pathak, D., Krahenbuhl, P., Donahue, J., Darrell, T., Efros, A.A.: Context encoders: feature learning by inpainting. In: Proceedings of the IEEE Conference on Computer Vision and Pattern Recognition, pp. 2536–2544 (2016)
15. Dolhansky, B., Ferrer, C.C.: Eye in-painting with exemplar generative adversarial networks. In: Proceedings of the IEEE Conference on Computer Vision and Pattern Recognition, pp. 7902–7911 (2018)
16. Yang, C., Lu, X., Lin, Z., Shechtman, E., Wang, O., Li, H.: High-resolution image inpainting using multi-scale neural patch synthesis. In: Proceedings of the IEEE Conference on Computer Vision and Pattern Recognition, pp. 6721–6729 (2017)
17. Iizuka, S., Simo-Serra, E., Ishikawa, H.: Globally and locally consistent image completion. ACM Trans. Graph. (ToG) **36**(4), 1–14 (2017)
18. Yu, J., Lin, Z., Yang, J., Shen, X., Lu, X., Huang, T.S.: Generative image inpainting with contextual attention. In: Proceedings of the IEEE Conference on Computer Vision and Pattern Recognition, pp. 5505–5514 (2018)
19. Cheng, J., Jaiswal, A., Wu, Y., Natarajan, P., Natarajan, P.: Style-aware normalized loss for improving arbitrary style transfer. In: Proceedings of the IEEE/CVF Conference on Computer Vision and Pattern Recognition, pp. 134–143 (2021)
20. Paysan, P., Knothe, R., Amberg, B., Romdhani, S., Vetter, T.: A 3D face model for pose and illumination invariant face recognition. In: 2009 Sixth IEEE International Conference on Advanced Video and Signal Based Surveillance, pp. 296–301. IEEE (2009)
21. Ramamoorthi, R., Hanrahan, P.: An efficient representation for irradiance environment maps. In: Proceedings of the 28th Annual Conference on Computer Graphics and Interactive Techniques, pp. 497–500 (2001)
22. He, K., Zhang, X., Ren, S., Sun, J.: Deep residual learning for image recognition. In: Proceedings of the IEEE Conference on Computer Vision and Pattern Recognition, pp. 770–778 (2016)

23. Schroff, F., Kalenichenko, D., Philbin, J.: FaceNet: a unified embedding for face recognition and clustering. In: Proceedings of the IEEE Conference on Computer Vision and Pattern Recognition, pp. 815–823 (2015)
24. Chen, A., Chen, Z., Zhang, G., Mitchell, K., Yu, J.: Photo-realistic facial details synthesis from single image. In: Proceedings of the IEEE International Conference on Computer Vision, pp. 9429–9439 (2019)
25. Bagdanov, A.D., Del Bimbo, A., Masi, I.: The florence 2D/3D hybrid face dataset. In: Proceedings of the 2011 Joint ACM Workshop on Human Gesture and Behavior Understanding, pp. 79–80 (2011)
26. Hernandez, M., Hassner, T., Choi, J., Medioni, G.: Accurate 3D face reconstruction via prior constrained structure from motion. Comput. Graph. **66**, 14–22 (2017)
27. Wang, Y., et al.: Look at the sky: sky-aware efficient 3D Gaussian splatting in the wild. IEEE Trans. Vis. Comput. Graph. (2025)
28. Zhang, Y., Zhang, H., Wu, G., Li, J.: Spatio-temporal self-supervision enhanced transformer networks for action recognition. In: 2022 IEEE International Conference on Multimedia and Expo (ICME), pp. 1–6. IEEE (2022)
29. Wang, Y., Wang, J., Wang, C., Duan, W., Bao, Y., Qi, Y.: SCARF: scalable continual learning framework for memory-efficient multiple neural radiance fields. In: Computer Graphics Forum, vol. 43, p. e15255. Wiley (2024)

Unified Multi-representation Modeling for Active Neural Reconstruction

Shuaixian Wang, Yaokun Li, Chenhui Guo, and Guang Tan(✉)

School of Intelligent Systems Engineering, Sun Yat-sen University, Shenzhen 510275, Guangdong, China
{wangshx29,liyk58,guochh25}@mail2.sysu.edu.cn, tanguang@mail.sysu.edu.cn

Abstract. We present a unified multi-representation modeling framework for neural active reconstruction that overcomes the limitations of single-feature approaches in complex environments. Our system innovatively combines occupancy fields, signed distance functions, and neural radiance fields within a hierarchical framework to achieve high-fidelity 3D reconstruction. The proposed method features: (1) a hybrid representation architecture that dynamically integrates discrete and continuous features, and (2) an active perception mechanism that optimizes viewpoint selection based on representation uncertainty. Comprehensive evaluations on Replica and MP3D datasets demonstrate significant improvements over existing methods. This work establishes a new paradigm for active neural reconstruction by bridging traditional SLAM with modern neural implicit representations.

Keywords: Hybrid Representations · Volume Rendering, Reconstruction

1 Introduction

Visual Simultaneous Localization and Mapping (SLAM) is a key task in computer vision, vital for applications like autonomous driving and robotic navigation [1]. It constructs maps of unknown environments using sequential spatial data, while simultaneously enabling the robot positioning.

Traditional approaches like Active-SLAM and Active-Map mainly rely on explicit 3D representations. Recently, advances in implicit representations represented by neural radiance fields (NeRF) have shown promise in diverse graphic applications, such as surface reconstruction [3] and novel view rendering [9]. These developments also provide new perspectives on active reconstruction. For instance, NeurAR [6] focuses on defining and learning a generalizable metric in the implicit representation to evaluate candidate viewpoints.

Despite the progress made, these approaches typically focus on collecting large amounts of data without adequately managing the inherent uncertainty associated with different representations. In more complex environments, where variations in lighting, texture, or viewpoint may introduce ambiguity into the

A. Ito (Ed.): ICMSSP 2025, LNNS 1637, pp. 21–28, 2025.
https://doi.org/10.1007/978-3-032-05994-9_3

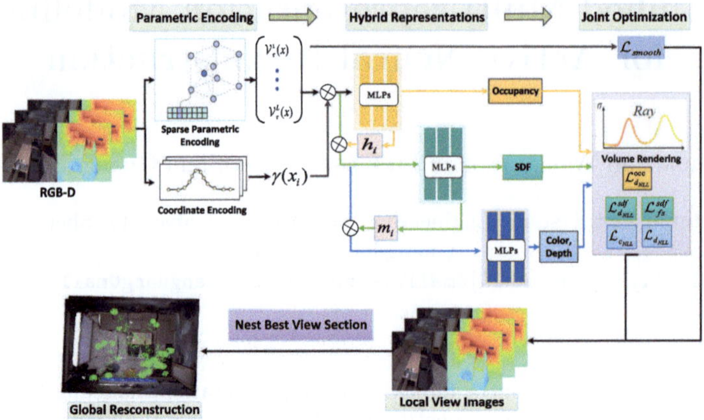

Fig. 1. The overview of our proposed architecture.

data, failing to address uncertainty within implicit representations can lead to less accurate reconstructions. This can result in issues such as poor hole filling, surface incoherence, and incompleteness in unobserved areas.

2 Methodology

2.1 The Overview of Architecture

Neural implicit representations enabled the encoding of both the appearance and 3D geometry of a scene. Given a stream of observations (RGB-D images) with known camera intrinsics, scene reconstruction can be achieved by optimizing neural scene representations using various approaches, such as radiance fields, occupancy maps, or truncated signed distance functions (TSDF). However, the study by [4,10] indicates that sole SDF-based neural surface representations hinder the implementation of scene-level geometric supervision signals, resulting in poor reconstructions of small and thin structures. On the other hand, occupancy-based representations tend to generate excessive structures but fail to ensure smooth surface reconstruction. Building on this insight, we propose hybrid representations that combine signed distance functions (SDFs) with occupancy representations to optimize the reconstruction process.

Moreover, to overcome the challenges of slow convergence and catastrophic forgetting in continual learning scenarios, we follow the study of CO-SLAM [8] and employ joint coordinate and parametric encoding. This approach enhances fidelity and expedites training by incorporating one-blob coordinate encoding with a multi-resolution hash-based feature grid. Based on the above fundamental hybrid scene representations and initial encoding, we design the hierarchical learning architecture, as shown in Fig. 1. The input RGBD images are processed into hash-based feature grid $\mathcal{V}_\tau = \{\mathcal{V}_\tau^l\}_{l=1}^L$ and feature vectors $\mathcal{V}_\tau(\boldsymbol{x})$ by sparse parametric encoding and coordinates encoding, then the initial representations are processed by MLPs networks into occupancy, SDF and color representations. We use joint optimization in the rendering processing for novel view synthesis and active reconstruction.

2.2 Hybrid Representations for Neural Reconstruction

Specially, our implicit representation maps the world coordinates $\boldsymbol{x}$ into color $\boldsymbol{c}$, truncated signed distance (TSDF) s and occupancy o values:

$$f_\theta(\mathbf{x}) \mapsto (\boldsymbol{c}, s, o) \tag{1}$$

We use one-blob encoding the spatial coordinates $\gamma(\boldsymbol{x})$. As the initial scene representation, we adopt a multi-resolution hash-based feature grid $\mathcal{V}_\tau = \{\mathcal{V}_\tau^l\}_{l=1}^L$. Feature vectors $\mathcal{V}_\tau(\boldsymbol{x})$ at each sampled point $\boldsymbol{x}$ are queried via trilinear interpolation. The scheme of our hierarchical hybrid learning representations is illustrated as follows: first, the geometry SDF MLP decoder f_α outputs the predicted SDF value s and a feature vector $\boldsymbol{h}$:

$$f_\alpha(\gamma(\boldsymbol{x}), \mathcal{V}_\tau(\boldsymbol{x})) \mapsto (\boldsymbol{h}, s) \tag{2}$$

Then, the occupancy MLPs denoted as f_β predicts the occupancy value:

$$f_\beta(\gamma(\boldsymbol{x}), \boldsymbol{h}) \mapsto (\boldsymbol{m}, o) \tag{3}$$

Finally, the color MLPs denoted as f_ϕ predicts the RGB value:

$$f_\phi(\gamma(\boldsymbol{x}), \boldsymbol{h}, \boldsymbol{m}) \mapsto \boldsymbol{c} \tag{4}$$

The parameters $\theta = \{\tau, \alpha, \beta, \phi\}$ are learnable and can be optimized during the bundle adjustment process.

For color rendering, according to [5], the color for the current ray r is rendered by

$$\hat{C}(r) = \sum_{i=1}^{M} T_i^r \alpha_i \hat{c}_i^r \tag{5}$$

where T_i^r and α_i represent the transmittance and alpha value, respectively, of the sampled point. They can be computed by

$$T_i^r = \prod_{j=1}^{i-1}(1-\alpha_i), \ \alpha_i = 1 - \exp(-\sigma_i^r \delta_i^r) \tag{6}$$

where δ_r^i is the distance between adjacent sample points.

For signed distance function (SDF) representation, we use the continuous function f_α that calculates the distance between each point and its closest surface. Following NeRF, the color of the current ray is computed via numerical integration, thus we can rewrite the neural rendering formula to

$$\hat{C}_{sdf}(r) = \sum_{i=1}^{N} f_\alpha(\mathbf{p}_i) \prod_{j<i}(1 - f_\alpha(\mathbf{x}_j)) c(\mathbf{p}_i, d) \tag{7}$$

In contrast to the SDF representation, occupancy modeling considers each point individually, independent of interactions with other objects in this complex scenario. In addition, the occupancy representation assumes that objects are solid, thus we can rewrite the neural rendering formula to

$$\hat{C}_{occ}(r) = \sum_{i=1}^{N} f_\beta(\mathbf{p}_i) \prod_{j<i} (1 - f_\beta(\mathbf{x}_j)) c(\mathbf{p}_i, d) \tag{8}$$

which replaces the opacity α to a discrete occupancy indicator variable $f_\beta \in 0, 1$, where $f_\beta = 0$ indicates the free space while $f_\beta = 1$ the occupied space. The function $f_\beta(\mathbf{p}_i)$ signifies the existence of an object at $\mathbf{p}_i$, a specific point on the object's surface. Meanwhile, $\mathbf{x}_j$ is a point along the course of a light ray.

2.3 Joint Volumetric Rendering Optimization

We use volumetric rendering optimization based on hierarchical uncertainty for bundle adjustment. During this process, 3D points are sampled along a ray, given a camera pose $\mathbf{o}$ and a ray direction $\mathbf{r}$, at predefined depths d_i: $\mathbf{x}_i = \mathbf{o} + d_i\mathbf{r}$. The rendered color $\hat{\mathbf{c}}$ and depth $\hat{\mathbf{d}}$ can be described as:

$$\hat{\mathbf{c}} = \frac{1}{\sum_{i=1}^{N} w_i} \sum_{i=1}^{N} w_i \mathbf{c}_i, \quad \hat{d} = \frac{1}{\sum_{i=1}^{N} w_i} \sum_{i=1}^{N} w_i d_i \tag{9}$$

where $w_i = \varphi\left(\frac{s_i}{tr}\right) \varphi\left(-\frac{s_i}{tr}\right)$ represents the weights calculated along the ray. We minimize a multi-objective function in the following bundle adjustment, incorporating color, depth, SDF, and occupancy rendering loss. These losses are derived from the comparison between the rendered outputs $(\hat{\mathbf{c}}, \hat{d})$ and the corresponding observed color and depth $(\mathbf{c}, D)$:

$$\begin{aligned} \mathcal{L}_c &= \frac{1}{N} \sum_{i=1}^{N} (\hat{\mathbf{c}}_i - \mathbf{c}_i)^2, \\ \mathcal{L}_d &= \frac{1}{|R_d|} \sum_{r \in R_d} (\hat{d}_r - D_r)^2 \end{aligned} \tag{10}$$

In this context, N and R_d represent the collection of rays that possess valid depth values, and D_r signifies the specific pixel located on the image.

In addition, for samples within the truncation region S_r^{tr}, SDF loss approximates the discrepancy between the sampled point and its observed depth value.

$$\mathcal{L}_{depth}^{sdf} = \frac{1}{|R_d|} \sum_{r \in R_d} \frac{1}{|S_r^{tr}|} \sum_{p \in S_r^{tr}} (s_p - (D_p - d))^2 \tag{11}$$

However, when it comes to points that are out of the truncation area, denoted by S_r^{fs}, a special requirement known as a free-space loss assures that the SDF estimates are fixed to tr.

$$\mathcal{L}_{fs}^{sdf} = \frac{1}{|R_d|} \sum_{r \in R_d} \frac{1}{|S_r^{fs}|} \sum_{p \in S_r^{fs}} (s_p - tr)^2 \tag{12}$$

Following the occupancy representation neural rendering formula as shown in (8), the depth rendering loss supervised by occupancy can be defined as Eq. 13. $\mathbf{occ} \sim \mathcal{N}\left(\hat{\mathbf{occ}}, \hat{\sigma}_{\mathbf{occ}}(\mathbf{r})^2\right)$ Then the predicted occupancy depth can be modified as Eq. ??:

$$\mathcal{L}_{depth}^{occ} = \sum_{r \in R} ||(w\hat{D}(r) + q) - \bar{D}(r)||^2 \tag{13}$$

To maintain the continuity of reconstructions in areas of free space where observation hasn't occurred, we execute a smoothness regularization on the interpolated features $V_\tau(\mathbf{x})$:

$$\mathcal{L}_{smooth} = \frac{1}{|\mathcal{G}|} \sum_{\mathbf{x} \in \mathcal{G}} \Delta_x^2 + \Delta_y^2 + \Delta_z^2 \tag{14}$$

where $\Delta_{x,y,z} = V_\tau(\mathbf{x}+\delta_{x,y,z}) - V_\tau(\mathbf{x})$ is the feature difference of adjacent sampled vertices on the hash grids along the three dimensions (Fig. 2).

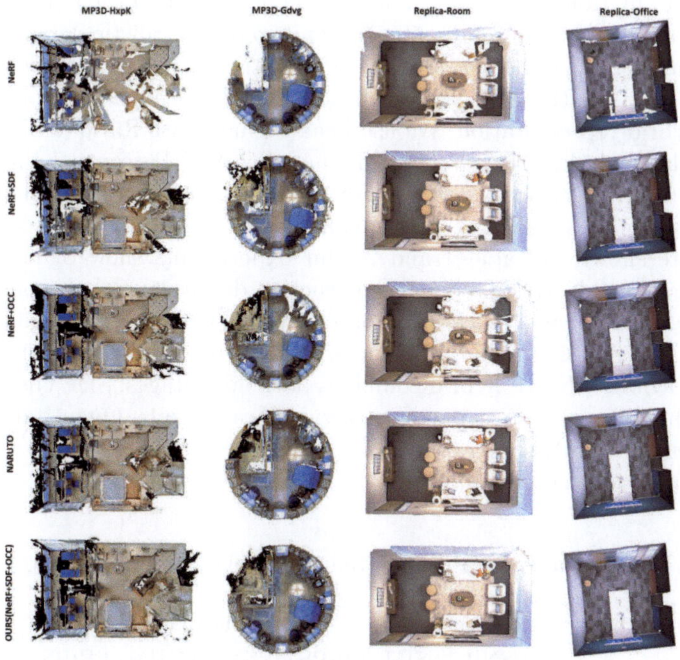

Fig. 2. Qualitative Results on Replica and MP3D Datasets

Finally, the joint loss is as Eq. 15, where $\lambda_1, \lambda_2, \lambda_3, \lambda_4$ are represented by uncertainty weighting from the uncertainty MLP network during training.

$$\mathcal{L} = \mathcal{L}_c + \mathcal{L}_d + \lambda_1 \mathcal{L}_d^{sdf} + \lambda_2 \mathcal{L}_d^{occ} + \lambda_3 \mathcal{L}_{fs}^{sdf} + \lambda_4 \mathcal{L}_{smooth} \tag{15}$$

3 Experiments

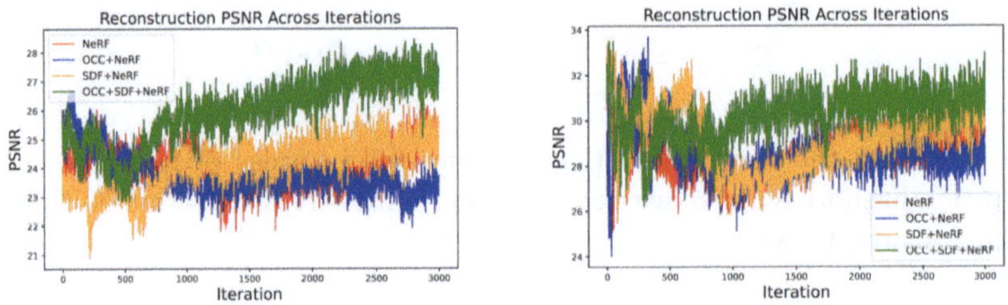

(a) Gdvg-PSNR (b) Room-PSNR

Fig. 3. Comparison of reconstruction PSNR across iterations.

We conduct ablation studies to evaluate the contribution of the individual representations and the hybrid representations on two photorealistic datasets: Replica [7] and Matterport3D (MP3D) [2]. We compare the reconstruction results among the NeRF only, NeRF+SDF representation, NeRF+Occupancy representation, and our hybrid representation. We use Peak Signal Noise Ratio (PSNR), Structural Similarity Index Measure (SSIM), and Learned Perceptual Image Patch Similarity (LPIPS), which leverages perceptual image similarity through insights derived from learned features. The quantitative results of Replica and MP3D datasets are shown in Tables 1 and 2, Our experiments show SDF-NeRF surpasses Occupancy-NeRF in scene reconstruction due to its global geometric constraints, which reduce artifacts like floaters. While occupancy methods excel at capturing fine details in thin structures, they struggle with local inconsistencies. Our Occ-SDF-NeRF hybrid combines both advantages: SDF ensures global surface coherence while occupancy preserves local details. Quantitative results (Fig. 3) demonstrate our hierarchical learning scheme achieves better optimization than individual representations.

Table 1. Ablation Study on Replica Results.

Method	Replica-Room			Replica-Office		
	PSNR ↑	SSIM↑	LPIPS ↓	PSNR ↑	SSIM↑	LPIPS ↓
NeRF	28.29	0.967	0.219	29.20	0.972	0.182
NeRF+SDF	29.71	0.974	0.213	29.53	0.922	0.147
NeRF+OCC	28.74	0.968	0.197	30.23	0.941	0.174
Ours	**30.46**	**0.979**	**0.182**	**30.76**	**0.978**	**0.171**

Table 2. Ablation study MP3D Results.

Method	MP3D- Hxpk			MP3D-Gdvg		
	PSNR ↑	SSIM↑	LPIPS ↓	PSNR ↑	SSIM↑	LPIPS ↓
NeRF	23.49	0.920	0.243	24.29	0.937	0.217
NeRF+SDF	24.39	0.929	0.213	26.30	0.954	0.198
NeRF+OCC	23.65	0.925	0.238	25.08	0.941	0.246
Ours	**26.67**	**0.961**	**0.136**	**27.03**	**0.964**	**0.185**

4 Conclusion

This paper presents a novel neural active reconstruction system that leverages hierarchical learning from hybrid representations. By integrating occupancy and signed distance functions into neural radiance fields, we enable joint volume rendering and multi-level scene understanding. Extensive experiments on benchmark datasets such as Replica and MP3D demonstrate that our approach outperforms existing methods in both qualitative and quantitative evaluations, achieving high reconstruction accuracy and completeness in simulated active exploration scenarios.

References

1. Cadena, C., et al.: Past, present, and future of simultaneous localization and mapping: toward the robust-perception age. IEEE Trans. Robot. **32**(6), 1309–1332 (2016). https://doi.org/10.1109/tro.2016.2624754
2. Chang, A., et al.: Matterport3d: learning from RGB-D data in indoor environments. In: 2017 International Conference on 3D Vision (3DV), pp. 667–676. IEEE (2017). https://doi.org/10.1109/3dv.2017.00081
3. Li, Y., Wang, S., Tan, G.: Id-nerf: indirect diffusion-guided neural radiance fields for generalizable view synthesis. Expert Syst. Appl. 126068 (2024). https://doi.org/10.1016/j.eswa.2024.126068
4. Lyu, X., et al.: Learning a room with the OCC-SDF hybrid: signed distance function mingled with occupancy aids scene representation. In: Proceedings of the IEEE/CVF International Conference on Computer Vision, pp. 8940–8950 (2023). https://doi.org/10.1109/iccv51070.2023.00821
5. Mildenhall, B., Srinivasan, P.P., Tancik, M., Barron, J.T., Ramamoorthi, R., Ng, R.: NERF: representing scenes as neural radiance fields for view synthesis. Commun. ACM **65**(1), 99–106 (2021). https://doi.org/10.1145/3503250
6. Ran, Y., et al.: Neurar: neural uncertainty for autonomous 3D reconstruction with implicit neural representations. IEEE Robot. Autom. Lett. **8**(2), 1125–1132 (2023). https://doi.org/10.1109/lra.2023.3235686
7. Straub, J., et al.: The replica dataset: a digital replica of indoor spaces. arXiv preprint arXiv:1906.05797 (2019)
8. Wang, H., Wang, J., Agapito, L.: Co-slam: joint coordinate and sparse parametric encodings for neural real-time slam. In: Proceedings of the IEEE/CVF Conference on Computer Vision and Pattern Recognition, pp. 13293–13302 (2023). https://doi.org/10.1109/cvpr52729.2023.01277

9. Wang, S., Haoran, X., Li, Y., Chen, J., Tan, G.: IE-NERF: exploring transient mask inpainting to enhance neural radiance fields in the wild. Neurocomputing **618**, 129112 (2025)
10. Haoran, X., et al.: Exploiting continuous motion clues for vision-based occupancy prediction. In: Proceedings of the AAAI Conference on Artificial Intelligence, vol. 39, pp. 8860–8868 (2025)

Geometry-Aware Face Reconstruction Under Occluded Scenes

Dapeng Zhao and Xiaoran Yan(✉)

Zhejiang Lab, Hangzhou 311121, China
yanxr@zhejianglab.com

Abstract. Recently, deep learning-based 3D face reconstruction methods have demonstrated promising advancements in terms of quality and efficiency. Nevertheless, these techniques face challenges in effectively handling occluded scenes and fail to capture intricate geometric facial details. Inspired by the principles of GANs and bump mapping, we have successfully addressed these issues. Our approach aims to deliver comprehensive 3D facial reconstructions, even in the presence of occlusions. While maintaining the overall shape's robustness, we introduce a mid-level shape refinement to the fundamental structure. Furthermore, we illustrate how our method adeptly extends to generate plausible details for obscured facial regions. We offer numerous examples that showcase the effectiveness of our framework in producing realistic results, where traditional methods often struggle. To substantiate the superior adaptability of our approach, we have conducted extensive experiments in the context of general 3D face reconstruction tasks, serving as concrete evidence of its regulatory prowess compared to manual occlusion removal methods.

Keywords: 3D Face Reconstruction · Face Alignment · Face Parsing · Occluded Scenes

1 Introduction

In the digital economy era, deriving a human face's shape from a single image is a cutting-edge area of research [1]. As individuals seek an improved quality of life, the challenge of face recognition in obstructed settings has gained significant attention, closely tied to 3D face reconstruction. Yet, many recent reconstruction methods prioritize texture accuracy, often overlooking extreme scenarios.

Thanks to advancements in deep learning, certain techniques can yield intricate 3D facial shapes [2, 3]. However, when faces are partially occluded, these methods tend to either reconstruct occlusions indiscriminately or fail outright. In previous studies, scholars have typically relied on the Shape from Shading (SfS) method [4, 5] to capture the geometric details of faces or objects for the purpose of three-dimensional shape reconstruction. The SfS approach infers depth information about an object's surface by analyzing variations in illumination and shadows within an image, thereby reconstructing its geometric structure. However, this method exhibits certain limitations in practical

A. Ito (Ed.): ICMSSP 2025, LNNS 1637, pp. 29–39, 2025.
https://doi.org/10.1007/978-3-032-05994-9_4

applications, primarily due to its dependence on prior knowledge regarding the target object's geometry or lighting conditions to effectively address the inherent ambiguities in the three-dimensional reconstruction process. Such ambiguities often arise from the complex interplay of multiple factors, including illumination, material reflectance properties, and viewing angles. Furthermore, when the input image fails to meet the assumptions underlying the SfS method—such as uniform lighting, the absence of complex shadows, or surface materials adhering to the Lambertian reflectance model—the reconstructed results may exhibit significant biases or inaccuracies. Consequently, these challenges constrain the method's applicability and robustness under non-ideal conditions.

In most cases, general face reconstruction methods often fail to provide geometric details [6, 7] or typically provide few details [7] to avoid reconstruction failures in extreme, challenging, unconstrained scenes. Most of these methods are based on localizing facial landmarks [8–10]. Unlike previous work, we describe an approach designed to attain both goals: Detailed 3D face reconstruction and robustness to occluded conditions.

The Main Contributions are Summarized as Follows:

- We propose an algorithm that combines facial landmarks and the face parsing map to identify the occluded region.
- To reconstruct the occluded area, we design a synthesis subnet that employs the predicted landmarks as guidance.
- We propose a novel 3D face reconstruction method that can produce reasonable face geometry details under occluded scenes.

2 Related Arts

2.1 3D Morphable Model

With the advancement of deep learning, significant progress has been made in 3D reconstruction, especially in the context of face reconstruction, utilizing Convolutional Neural Networks. The method proposed by Anh *et al.* [11] offers both dense face alignment and facial model results. However, the effectiveness of these approaches is constrained by the limitations inherent in the 3D space defined by templates.

In contrast, some recent end-to-end techniques [9, 12] have achieved state-of-the-art performance in their respective domains. These methods, however, typically directly reconstruct the facial model without the need for input image identification or processing. VRN [9] introduces an innovative deep 3D face reconstruction approach that eliminates the necessity for precise alignment and the establishment of dense image correspondences, requiring only a single 2D facial image. VRN significantly widens the range of applications for 3D face technology. On the other hand, PRNet [12] has devised a straightforward method for simultaneously reconstructing the 3D facial structure. This approach incorporates a UV position map to record the 3D facial shape in UV space.

2.2 Generative Adversarial Nets

Under normal circumstances, we believe that Generative Adversarial Nets (GANs) are among the few tools capable of achieving a reconstruction effect in the 2D face field. After identifying and removing the occluded areas, we employ GAN to synthesize the facial image while preserving a reasonable topological structure. GAN typically comprises a generator and a discriminator, which engage in a competitive process. In the realm of 2D images, GAN stands out as one of the most effective tools for face synthesis. GAN has found successful applications in the fields of image manipulation [13–16] and image-to-image translation [17–21].

2.3 Facial Landmark Detection

The positions of facial landmark points, distributed around key facial components and the facial contour, precisely capture both rigid and non-rigid facial deformations induced by head movements and facial expressions. These landmarks play an indispensable role in documenting facial geometry and dynamic variations, thus holding critical value in three-dimensional facial reconstruction tasks, particularly for achieving high-precision texture reconstruction. Accurate landmark localization not only facilitates the construction of more realistic three-dimensional facial models but also provides reliable foundational data for subsequent texture mapping and detail optimization. Over the years, extensive research in academia and industry has focused on facial landmark detection, leading to the development of various algorithms for the automated identification of these critical points. These algorithms differ in their theoretical foundations and application contexts, significantly contributing to advancements in facial analysis. Traditional facial landmark detection methods predominantly rely on regression-based approaches, such as predicting landmark positions through regression models, which typically depend on the extraction and analysis of local features, integrating statistical models or machine learning techniques to iteratively approximate the true positions of landmarks, thereby achieving relatively precise detection outcomes. Valstar *et al.* [22] predict landmark locations from face image patches using support vector regression (SVR). Cao *et al.* [23] use cascaded regression with pixel-difference features. Several other works [24–26] employ random regression forests to cast votes for landmark locations based on local face images. In these methods, an initial landmark position is first randomly selected, and then the facial landmarks are iteratively predicted. Therefore, the quality of the initial position is crucial.

3 Our Method

3.1 Landmark Prediction Network

Face alignment (especially accurate facial landmark $\mathbf{L_{face}} \in \mathbb{R}^{2\times 68}$ generation) is the basis of the whole framework (Section 1 of Fig. 1). The closest approach [27] to our method is MobileNet-V3model [28]. The network $\mathcal{N}_{pre}$ aims to generate $\mathbf{L_{face}}$ from a face image $\mathbf{I_0} : \mathbf{L_{face}} = \mathcal{N}_L(\mathbf{I_0}; \theta_{land})$, where θ_{land} denotes the parameters to be trained. Since our task is to export face features rather than identify different people, we design the loss function as follows:

$$\mathcal{L}_{lmk} = \left\| \mathbf{L_{face}} - \mathbf{L_{gt}} \right\|_2^2$$

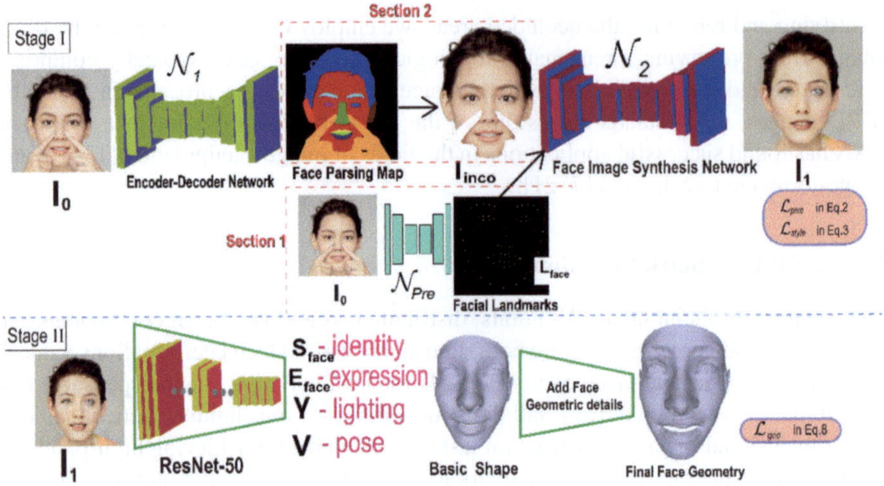

Fig. 1. Our method overview.

where $\mathbf{L_{gt}}$ is the ground truth face landmarks and $\|\cdot\|_2$ is the L_2 norm.

As shown in Fig. 1, our method is divided into two phases: the final output of stage 1 is to remove the 2D face picture of the occlusion; the final output of stage 2 is a 3D face model with geometric details.

3.2 Face Synthesis Network

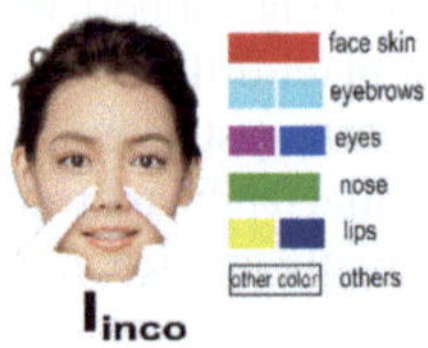

Fig. 2. Face parsing network of our method.

Overall, We design the image synthesis module $\mathcal{N}_s$ to synthesize a 2D face that removes the occlusion region. The module includes three parts: matting operator (Section 2 of Fig. 1), generator and discriminator.

Matting Operator. Typically, the task of the Matting operator $\mathcal{N}_1$ is to remove the occluded areas $\mathbf{I_m}$ of the input photo $\mathbf{I_0}$ (Fig. 2). For each input face image $\mathbf{I_{in}} \in \mathbb{R}^{H\times W\times 3}$, we utilized the trained model $\mathcal{N}_1$ to estimate pixel-level label classes and generate the face parsing map $\mathbf{Q} \in \mathbb{R}^{H\times W\times 1}$. According to the map $\mathbf{Q}$, we identify and remove the occluded area $\mathbf{I_m}$ to obtain the hollowed-out photo $\mathbf{I_{inco}}$. This will be the input for our next task.

Generator. The generator $\mathcal{N}_2$ desires to generate the complete face by taking hollowed-out images $\mathbf{I_{inco}}$ and landmarks $\mathbf{L}$ ($\mathbf{L_{face}}$ or $\mathbf{L_{gt}}$). More specifically, the network includes gradually down-sampled encoding blocks. These blocks are composed of subordinate residual blocks with dilated convolutions and an attention block. The generator can be formulated as $\mathbf{I_1} = \mathcal{N}_2(\mathbf{I_{inco}}, L; \theta_2)$, with θ_2 the trainable parameters.

Discriminator. The purpose of the discriminator is to judge whether the data distribution meets our requirements. The ambition of face synthesis is achieved when the generated results are not distinguishable from the real ones.

Loss Function. We use a combination of per-pixel loss and style loss for training the face synthesis network.

We compute the per-pixel loss as follows:

$$\mathcal{L}_{pixe} = \frac{1}{M_s}\|\mathbf{I_1} - \mathbf{I_0}\|$$

where M_s denotes the mask size and $\|\cdot\|$ stands for the L_1 norm. Notice that we use M_s as the denominator and its role is to adjust the penalty.

The style loss computes the style distance between two images as follows:

$$\mathcal{L}_{style} = \sum_{\mathrm{n}} \frac{1}{R_n \times R_n}\left\| \frac{B_n(\mathbf{I_1} \odot \mathbf{I_m}) - G_n(\mathbf{I_0} \odot \mathbf{I_m})}{R_n \times H_n \times W_n} \right\|$$

where $B_{\mathrm{n}}(\mathrm{x}) = \varphi_n(x)^T \varphi_n(x)$ denotes the Gram Matrix corresponding to $\varphi_n(x)$, $\varphi_n(\cdot)$ stands for the R_n feature maps with the size of the $H_n \times W_n$ n-th layer.

The total loss with respect to the face synthesis module:

$$\mathcal{L}_{final} = \lambda_{pixe}\mathcal{L}_{pixe} + \lambda_{style}\mathcal{L}_{style}$$

here, we use $\lambda_{pixe} = 1$, $\lambda_{style} = 250$ in our experiments.

3.3 Construct the Fundamental Shapes

We utilize the hot deep learning method [29] in recent years to regress the 3DMM coefficients to construct the basic face shape. The primary face model is formulated as follows:

$$\mathbf{S_{face}} = \overline{\mathbf{S}} + \sum_{i=1}^{m-1} \alpha_i \mathbf{s_i}$$

where $\overline{\mathbf{S}}$ denotes an average 3D face shape, m denotes the number of faces participating in the weighted face datasets, $\mathbf{s_i}$ denotes the ith face shape and α_i denotes the face shape coefficients estimated from $\mathbf{I_1}$.

Similarly, we model the expressions using the following formulation:

$$\mathbf{E_{face}} = \sum_{j=1}^{m-1} \beta_i \mathbf{e_i}$$

where β_i denotes the expressions coefficients and $\mathbf{e_i}$ denotes the ith expression.

Further, we denote the viewpoint as $\mathbf{v}=[\mathbf{r}^T, \mathbf{t}^T]$, where $\mathbf{r} \in \mathbb{R}^3$ denotes the 3D rotation, expressed by Euler angles, and $\mathbf{t} \in \mathbb{R}^3$ denotes a translation vector, together aligning a generic 3D face shape with the face appearing in $\mathbf{I_1}$.

We adopt the Basel Face Model (BFM) [30], which provides both $\overline{\mathbf{S}}$, $\mathbf{s_i}$ and $\mathbf{e_i}$.We approximated the scene illumination with Spherical Harmonics (SH) [31–33] parameterized by coefficient vector $\boldsymbol{\gamma} \in \mathbb{R}^9$. Given $\mathbf{I_1}$, we estimate $y = (\mathbf{S_{face}}, \mathbf{E_{face}}, \mathbf{V}, \boldsymbol{\gamma})$using the recent deep 3DMM approach of Yu *et al.* [33], with their pre-trained model.

3.4 Recovering Face Geometric Details

Inspired by the method of image-to-image translation [17, 34], we define the displacements of the depth map as the distances through the pixels of $\mathbf{I_1}$ to the 3D face surface. Generally, we define the non-linear bump map $\Phi(\mathbf{b})$ as:

$$\Phi(\mathbf{b}) = \begin{cases} \phi(0) \text{ other} \\ \phi(d'(\mathbf{b}) - d(\mathbf{b})) \text{ face projects to } \mathbf{b} \end{cases}$$

where $\phi(\cdot)$ denotes an encoding function that converts the depth value to the linear range $[0, \ldots, 255]$, $\mathbf{b}$ denotes the pixel coordinate $\left[x, y\right]$ in $\mathbf{I_1}$, $d'(\mathbf{b})$ denotes the depth, which is the distance from the surface of the detailed face shape to $\mathbf{b}$ along the line of sight, $d(\mathbf{b})$ denotes the depth of the basic shape.

Thus, given a bump map Φ and the depth of the basic shape, we can compute the detailed depth follows $d'(\mathbf{b}) = d(\mathbf{b}) + \phi^{-1}(\Phi(\mathbf{b}))$.

In order to increase geometric details and to suppress noise, we define the loss function as follows:

$$\mathcal{L}_{geo} = \|\tilde{\Phi} - \Phi\| + \left\|\frac{\partial\tilde{\Phi}}{\partial x} - \frac{\partial\Phi}{\partial x}\right\| + \left\|\frac{\partial\tilde{\Phi}}{\partial y} - \frac{\partial\Phi}{\partial y}\right\|$$

where $\|\cdot\|$ denotes the L_1 norm, $\tilde{\Phi}$ denotes the ground truth and $\frac{\partial\tilde{\Phi}}{\partial x}, \frac{\partial\tilde{\Phi}}{\partial y}$ denotes the 2D gradient of the bump map. We found that by adding these last two terms of loss function and we reduce bump map noise by favoring smoother surfaces. At the same time, the final effect shows that high-frequency details are preserved.

4 Experimental Results

4.1 Qualitative Comparison with Other Methods

Figure 3 shows our reconstruction results compared with the contemporary arts [3, 6, 35]. It can be clearly seen from the Fig. 3 that Sela *et al.*'s method shows the shape of various obstructions. This may be due to the method focusing too much on the local shape (Fig. 4). Since the other two methods are based on existing 3D face model templates, the shapes are too smooth and lack geometric details. The template itself constitutes a simplified, averaged representation of a human face model. To reduce computational complexity, the triangular mesh of the template typically comprises a relatively limited number of

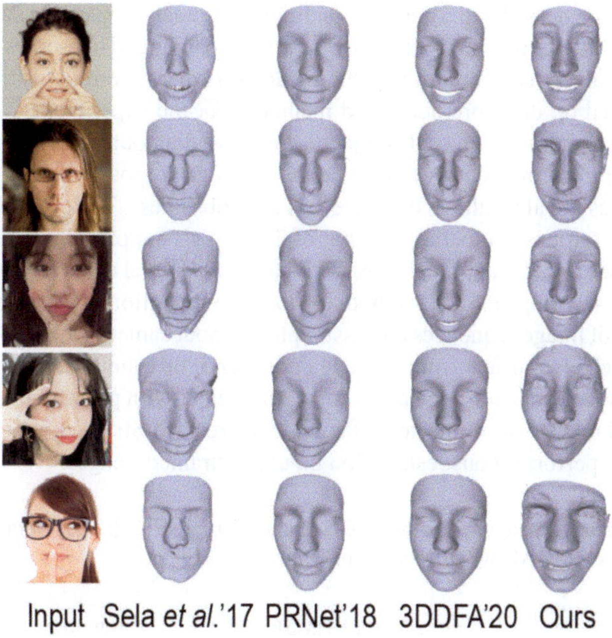

Fig. 3. Comparison of qualitative results. Baseline methods from left to right: Sela *et al.*, PRNet,3DDFA, and our method.

polygons, which constrains its capacity to represent high-frequency geometric details, such as wrinkles, pores, and scars. Parametric face models are generally grounded in a linear space, which is inherently incapable of fully capturing the nonlinear variations in facial shapes, thereby resulting in the loss of intricate details.

Fig. 4. Comparison of local facial geometric details.

4.2 Quantitative Comparison with Other Methods

To rigorously evaluate the superiority of our proposed method and demonstrate its robustness and adaptability on unconstrained face images, we used the MICC datasets [36] as the testing benchmark. We compared our method with the Structure-from-Motion (SFM) method [37, 38] and a learning-based method [11] through systematic experimental analysis. The MICC dataset includes 53 video sequences. Each sequence has different resolutions, shooting conditions, and scaling levels. Data collection occurred in controlled, semi-controlled, and fully uncontrolled environments to ensure diversity and representativeness of experimental conditions. For each subject, the dataset provides a ground truth model obtained via structured light scanning. This serves as the benchmark for evaluating reconstruction accuracy. Reconstruction errors of the results were calculated based on the aforementioned methods. For the experiment, we selected the most frontal face image from each subject's corresponding outdoor video. We used this image as input to reconstruct a 3D face model with our proposed method. This process further validates the algorithm's effectiveness and robustness. Table 1 indicates that our reconstruction error is comparable to that of Tran *et al.*'s approach [11] and surpasses that of Hernandez *et al.*'s approach [39]. This can be attributed to the fact that Hernandez *et al.*'s approach, being an SFM method, exhibits the limitation of excessive reliance on the continuity of image sequences and assumptions about camera motion. Although Tran *et al.*'s approach also suffers from a strong dependence on training data distribution and limited generalization to unseen data, its error rate remains on par with ours. Leveraging reliable prior knowledge of mid-level facial structures and SFS recovery, our approach can effectively perform robust estimation on unconstrained images.

Table 1. Quantitative Outcomes on the MICC Florence Datasets [36]. The Average and Standard Deviation of3DRMSE, Along with Execution Times.

Method	3DRMSE	run time
Hernandez *et al.*'s approach	1.92 ± 0.39	CPU 1 min 13 s
Tran *et al.*'s approach	1.53 ± 0.29	GPU 1.77 s
Ours	1.55 ± 0.35	CPU 1 min 55 s

5 Conclusions

We describe an approach capable of producing detailed 3D face reconstructions from images taken in occluded conditions. The advantage of our method is that it can effectively process the facial details in places where the face is occluded. Since the previous arts are limited to the reconstruction of faces in non-occluded scenes, our method represents a leap in 3D facial details reconstruction capabilities. Unlike the recently popular end-to-end reconstruction method, our method is a two-stage face reconstruction task based on basic shapes.

Acknowledgments. Supported by National Key R&D Program of China (No. 2022YFF0712400).

References

1. Vetter, T., Blanz, V.: Estimating coloured 3D face models from single images: an example based approach. In: Burkhardt, H., Neumann, B. (eds.) ECCV 1998. LNCS, vol. 1407, pp. 499–513. Springer, Heidelberg (1998). https://doi.org/10.1007/BFb0054761
2. Richardson, E., Sela, M., Or-El, R., Kimmel, R.: Learning detailed face reconstruction from a single image. In: Proceedings of the IEEE Conference on Computer Vision and Pattern Recognition, pp. 1259–1268 (2017)
3. Sela, M., Richardson, E., Kimmel, R.: Unrestricted facial geometry reconstruction using image-to-image translation. In: Proceedings of the IEEE International Conference on Computer Vision, pp. 1576–1585 (2017)
4. Kemelmacher-Shlizerman, I., Basri, R.: 3d face reconstruction from a single image using a single reference face shape. IEEE Trans. Pattern Anal. Mach. Intell. **33**(2), 394–405 (2010)
5. Li, C., Zhou, K., Lin, S.: Intrinsic face image decomposition with human face priors. In: Fleet, D., Pajdla, T., Schiele, B., Tuytelaars, T. (eds.) ECCV 2014. LNCS, vol. 8693, pp. 218–233. Springer, Cham (2014). https://doi.org/10.1007/978-3-319-10602-1_15
6. Guo, J., Zhu, X., Yang, Y., Yang, F., Lei, Z., Li, S.Z.: Towards fast, accurate and stable 3D dense face alignment. arXiv preprint arXiv:2009.09960 (2020)
7. Hassner, T.: Viewing real-world faces in 3D. In: Proceedings of the IEEE International Conference on Computer Vision, pp. 3607–3614 (2013)
8. Taigman, Y., Yang, M., Ranzato, M., Wolf, L.: DeepFace: closing the gap to human-level performance in face verification. In: Proceedings of the IEEE Conference on Computer Vision and Pattern Recognition, pp. 1701–1708 (2014)
9. Jackson, A.S., Bulat, A., Argyriou, V., Tzimiropoulos, G.: Large pose 3D face reconstruction from a single image via direct volumetric CNN regression. In: Proceedings of the IEEE International Conference on Computer Vision, pp. 1031–1039 (2017)
10. Zhu, X., Lei, Z., Liu, X., Shi, H., Li, S.Z.: Face alignment across large poses: a 3D solution. In: Proceedings of the IEEE Conference on Computer Vision and Pattern Recognition, pp. 146–155 (2016)
11. Tran, A.T., Hassner, T., Masi, I., Medioni, G.: Regressing robust and discriminative 3D morphable models with a very deep neural network. In: Proceedings of the IEEE Conference on Computer Vision and Pattern Recognition, pp. 5163–5172 (2017)
12. Feng, Y., Wu, F., Shao, X., Wang, Y., Zhou, X.: Joint 3D face reconstruction and dense alignment with position map regression network. In: Ferrari, V., Hebert, M., Sminchisescu, C., Weiss, Y. (eds.) ECCV 2018. LNCS, vol. 11218, pp. 557–574. Springer, Cham (2018). https://doi.org/10.1007/978-3-030-01264-9_33
13. Abdal, R., Qin, Y., Wonka, P.: Image2StyleGAN: how to embed images into the StyleGAN latent space? In: Proceedings of the IEEE/CVF International Conference on Computer Vision, pp. 4432–4441 (2019)
14. Jaderberg, M., Simonyan, K., Zisserman, A., Kavukcuoglu, K.: Spatial transformer networks. arXiv preprint arXiv:1506.02025 (2015)
15. Lee, C.-H., Liu, Z., Wu, L., Luo, P.: MaskGAN: towards diverse and interactive facial image manipulation. In: Proceedings of the IEEE/CVF Conference on Computer Vision and Pattern Recognition, pp. 5549–5558 (2020)
16. Zhu, J.-Y., Krähenbühl, P., Shechtman, E., Efros, A.A.: Generative visual manipulation on the natural image manifold. In: Leibe, B., Matas, J., Sebe, N., Welling, M. (eds.) ECCV 2016. LNCS, vol. 9909, pp. 597–613. Springer, Cham (2016). https://doi.org/10.1007/978-3-319-46454-1_36
17. Isola, P., Zhu, J.-Y., Zhou, T., Efros, A.A.: Image-to-image translation with conditional adversarial networks. In: Proceedings of the IEEE Conference on Computer Vision and Pattern Recognition, pp. 1125–1134 (2017)

18. Park, T., Liu, M.-Y., Wang, T.-C., Zhu, J.-Y.: Semantic image synthesis with spatially-adaptive normalization. In: Proceedings of the IEEE Conference on Computer Vision and Pattern Recognition, pp. 2337–2346 (2019)
19. Liu, M.-Y., Breuel, T., Kautz, J.: Unsupervised image-to-image translation networks. arXiv preprint arXiv:1703.00848 (2017)
20. Wang, T.-C., Liu, M.-Y., Zhu, J.-Y., Tao, A., Kautz, J., Catanzaro, B.: High-resolution image synthesis and semantic manipulation with conditional GANs. In: Proceedings of the IEEE Conference on computer Vision and Pattern Recognition, pp. 8798–8807 (2018)
21. Zhu, J.-Y., Park, T., Isola, P., Efros, A.A.: Unpaired image-to-image translation using cycle-consistent adversarial networks. In: Proceedings of the IEEE International Conference on Computer Vision, pp. 2223–2232 (2017)
22. Valstar, M., Martinez, B., Binefa, X., Pantic, M.: Facial point detection using boosted regression and graph models. In: 2010 IEEE Computer Society Conference on Computer Vision and Pattern Recognition, pp. 2729–2736. IEEE (2010)
23. Cao, X., Wei, Y., Wen, F., Sun, J.: Face alignment by explicit shape regression. Int. J. Comput. Vis. **107**(2), 177–190 (2014)
24. Cootes, T.F., Ionita, M.C., Lindner, C., Sauer, P.: Robust and accurate shape model fitting using random forest regression voting. In: Fitzgibbon, A., Lazebnik, S., Perona, P., Sato, Y., Schmid, C. (eds.) ECCV 2012. LNCS, vol. 7578, pp. 278–291. Springer, Heidelberg (2012). https://doi.org/10.1007/978-3-642-33786-4_21
25. Dantone, M., Gall, J., Fanelli, G., Van Gool, L.: Real-time facial feature detection using conditional regression forests. In: 2012 IEEE Conference on Computer Vision and Pattern Recognition, pp. 2578–2585. IEEE (2012)
26. Yang, H., Patras, I.: Sieving regression forest votes for facial feature detection in the wild. In: Proceedings of the IEEE International Conference on Computer Vision, pp. 1936–1943 (2013)
27. Zhang, Y., et al.: Bullet-screen-emoji attack with temporal difference noise for video action recognition. IEEE Trans. Circuits Syst. Video Technol. (2024)
28. Howard, A., et al.: Searching for MobileNetV3. In: Proceedings of the IEEE/CVF International Conference on Computer Vision, pp. 1314–1324 (2019)
29. Wang, Y., et al.: Look at the sky: sky-aware efficient 3D Gaussian splatting in the wild. IEEE Trans. Vis. Comput. Graph. (2025)
30. Paysan, P., Knothe, R., Amberg, B., Romdhani, S., Vetter, T.: A 3D face model for pose and illumination invariant face recognition. In: 2009 Sixth IEEE International Conference on Advanced Video and Signal Based Surveillance, pp. 296–301. IEEE (2009)
31. Tewari, A., Seidel, H.-P., Elgharib, M., Theobalt, C.: Learning complete 3D morphable face models from images and videos. arXiv preprint arXiv:2010.01679 (2020)
32. Deng, Y., Yang, J., Chen, D., Wen, F., Tong, X.: Disentangled and controllable face image generation via 3D imitative-contrastive learning. In: Proceedings of the IEEE/CVF Conference on Computer Vision and Pattern Recognition, pp. 5154–5163 (2020)
33. Deng, Y., Yang, J., Xu, S., Chen, D., Jia, Y., Tong, X.: Accurate 3D face reconstruction with weakly-supervised learning: from single image to image set. In: Proceedings of the IEEE Conference on Computer Vision and Pattern Recognition Workshops (2019)
34. Tran, A.T., Hassner, T., Masi, I., Paz, E., Nirkin, Y., Medioni, G.: Extreme 3D face reconstruction: seeing through occlusions. In: Proceedings of the IEEE Conference on Computer Vision and Pattern Recognition, pp. 3935–3944 (2018)
35. Feng, Y., Feng, H., Black, M.J., Bolkart, T.: Learning an animatable detailed 3D face model from in-the-wild images. ACM Trans. Graph. (TOG) **40**, 1–13 (2020)
36. Bagdanov, A.D., Del Bimbo, A., Masi, I.: The florence 2D/3D hybrid face dataset. In: Proceedings of the 2011 Joint ACM Workshop on Human Gesture and Behavior Understanding, pp. 79–80 (2011)

37. Zhang, Y., et al.: Temporal transformer networks with self-supervision for action recognition. IEEE Internet Things J. **10**(14), 12999–13011 (2023)
38. Wang, Y., Wang, J., Wang, C., Duan, W., Bao, Y., Qi, Y.: SCARF: scalable continual learning framework for memory-efficient multiple neural radiance fields. In: Computer Graphics Forum, vol. 43, p. e15255. Wiley (2024)
39. Hernandez, M., Hassner, T., Choi, J., Medioni, G.: Accurate 3D face reconstruction via prior constrained structure from motion. Comput. Graph. **66**, 14–22 (2017)

Image Analysis and Multimedia Application

A Simple yet Accurate Autoadaptive Model of Network Traffic for Detection of Attacks on Low Latency Services

Rémi Cogranne[1(✉)], Marius Letourneau[1], Guillaume Doyen[2], and Huu Nghia Nguyen[3]

[1] Troyes University of Technology, Troyes, France
{remi.cogranne,marius.letourneau}@utt.fr
[2] OCIF - IRISA (UMR CNRS 6074), IMT Atlantique, Rennes, France
guillaume.doyen@imt-atlantique.fr
[3] Montimage, Paris, France
huunghia.nguyen@montimage.eu

Abstract. This paper addresses the problem of detection of attacks in computer networks. More precisely, we consider attacks on emerging low-latency services, which typically require a specific traffic management system. We present a simple yet very efficient hybrid method that takes advantage of both autoencoders and transformer models. The original method is compared with the current state-of-the-art on a large real-life dataset of network traffic to show the relevance of the proposed approach, especially for low false-positive rates. A quick ablation analysis shows that the efficiency of the method relies on the combined use of the two approaches jointly in our hybrid model.

Keywords: Multivariate Time-Series · Network Security · Multidimensional Signal Processing · Attack detection · Cybersecurity

1 Introduction

The technological progression of network infrastructures such as the deployment of fibre-optic links, fifth-generation mobile networks (5G), Virtualized Network Functions (VNFs), and network slicing techniques, to cite a few, have opened an age of significantly elevated throughput rates and slashed latency periods [1,2]. These advancements have given rise to a variety of new applications with intrinsically low latency demands, such as cloud gaming, cloud robotics, tele-robotics, and tactile internet services. However, the development of novel Internet services opens new possible vectors of attacks and security must be a priority because, without any doubt, in today's digital age, the likelihood of an attack

M. Letourneau, G. Doyen and H. Nguyen—Contributing authors.

H.N. Nguyen—This work was partially supported by the French ANR MOSAICO project No. ANR-19-CE25-0012.

A. Ito (Ed.): ICMSSP 2025, LNNS 1637, pp. 43–58, 2025.
https://doi.org/10.1007/978-3-032-05994-9_5

increases along with the potential impact, and low latency services will become a target when deployed. In addition, as often seen in computer networks, Future Internet will not replace existing services, hence the birth of a coexistence of regular internet traffic, some with high-throughput targets, with such emerging latency-sensitive applications.

The Low Latency, Low Loss, and Scalable Throughput (L4S) architecture has been positioned as a prospective solution to meet these cutting-edge network demands. When tested, the L4S architecture demonstrated satisfying performance under standard traffic behaviour [3]. Yet, research lacks insight regarding its capability to work under abnormal or non-regulated traffic conditions. For instance, it has been shown in [4] how rather moderate traffic bursts can heavily impact L4S performance, thus preventing the delivery of intended Quality of Experience (QoE). This problem is even more concerning in the context of cyberattacks; for instance, it has been shown in [5] how it is possible to compromise cloud gaming platform services by leveraging so-called booters making a game become very unfair and ultimately compromising players' experience.

The present paper addresses the problem of the detection of attacks that specifically aims at disrupting low-latency services. As for a vast majority of similar DDoS detection problems, there are several major intrinsic difficulties. First the attacks on low-latency services are rather new; their means are not completely known and entirely characterized. Second, the traffic of a computer network is, in general, a complex signal that is difficult to analyse and model. Last and not least, the detection methodology must be reliable in the sense that it should be able to detect (1) attack with a limited delay, as quickly as possible and (2) it should prevent as much as possible false-alarm rate in order to avoid raising unwanted false alarms, which would undermine the overall system's usability.

We propose an original methodology based on an autoadaptive linear model for analysing network traffic measurements, considered in this paper as multivariate time series. Our model leverages the advantages of detection theory using a linear model, which is well established, and the relevance of unsupervised learning to build automatically an accurate model representing the multivariate time series under inspection.

In our prior works [6,7], we have described the impact of a specific attack so-called "unresponsive ECN" and assessed its practical efficiency throughout large-scale numerical evaluation and experimentation. Therefore, the detection method proposed in the present paper is evaluated on a large dataset of real network traffic data implemented in a realistic testbed with several legitimate clients and attackers.

We show that the proposed method, despite its simplicity, performs very well for multidimensional signal processing and detection.

The present paper is organized as follows. Section 2 provides a brief overview of low-latency network architectures currently being transferred from research labs to the real world. We recall the possible attack mechanisms on low-latency services and their impact on the traffic. Section 3 states the problem of dealing with nuisance parameters and presents the methodology used in the present work to represent and reject the legitimate traffic. Then Sect. 4 presents the autoadaptive linear model used to represent the multidimensional data that computer traffic networks are. It especially explains how unsupervised learning is used to adjust the proposed linear model with the previous observations in order to represent the traffic metrics accurately. Section 5 presents the proposed method for the detection of low-latency attacks. Section 6 assesses the proposed methodology on a large real-world dataset of low-latency attacks and legitimate traffic. Section 7 concludes the paper and draws possible future works.

2 Summary of Attacks on Low-Latency Services

In a nutshell, the L4S architecture rests on three main simple pillars. On the client endpoint side, L4S leverages accurate Explicit Congestion Notifications (ECN), which provides real-time indicators of network congestion to alert users such that they shall limit their traffic before a large congestion occurs. It also implements a novel congestion control algorithm (CCA), namely Prague, available both for the Transmission Control Protocol (TCP) and the Quick UDP Internet Connections (QUIC) protocol, to use as much as possible link bandwidth while avoiding traffic congestion [8,9]. On the network side, L4S uses a Dual-Queue Coupled Active Queue Management (AQM) system which has the ability to handle queueing priorities, coupled directly with a unique packet classifier that segregates signals for Low Latency (LL) and regular, legacy flows. This categorical discrimination is instrumental in a bid to institute targeted behaviour modification, allowing LL flows not to disproportionately influence the Classic bandwidth. The central performance of the dual AQM relies on this coupling mechanism, which shall find an equilibrium in moderating the classical queue, for legacy traffic, in order to preserve low-latency features for the second queue while preventing the low-latency flows' tendency to cause a so-called starvation of classic, legacy traffic [1,10].

In our previous work [6], we demonstrated that traffic that does not comply with the rules of ECN and TCP Prague protocols can significantly degrade the effective operation of L4S. Building on this, the present paper focuses on a specific type of malicious flow: "unresponsive yet ECN-compatible" flows. These flows pretend to adhere to ECN congestion notifications but fail to adjust their sending rates accordingly, thereby subverting the intended behaviour. It is worth noting that a small proportion of unresponsive traffic is inherent in computer network traffic because of UDP and VoIP communications, for instance. Fortunately, L4S is designed to accommodate a limited fraction of such flows, as acknowledged in [11]. However, as the proportion of unresponsive flows increases, L4S struggles to maintain both low-latency services and the throughput of regular flows. Alarmingly, even a tiny fraction of "unresponsive yet ECN-compatible" flows can rapidly degrade L4S performance and undermine any low-latency services.

We have also conducted large-scale experimentation in [7] to characterize the attack mechanism on L4S. Considering the metrics that a router can easily monitor, we showed that the impact of the attack is entirely characterized by the following seven measurements:

1. traffic rate, or number of bytes transmitted per second ;
2. in-router queueing delay, which is the average time a packer remains in the cache of the router before being retransmitted (for both low-latency and regular traffic) ;
3. the number of packets within both queues, which is the fraction of the total number of packets in the router's cache memory ;
4. the probability of ECN marking, which is the probability used by the dual queue system of setting ECN notification ;
5. probability of a packet drop, a packet is dropped when either the router drops it preventively or when its queue is fully occupied ;
6. the total number of ECN marks written on packets by the AQM, which can be due to natural queue-building behaviours within each queue that lead to an increase in the marking probability ;
7. the number of ECN marks that is due to a step threshold overflow of the low latency queue ;

It is especially interesting to note that a Principal Component Analysis shows that some of those metrics are extremely correlated under normal conditions ; yet it is shown in [7] that when a large fraction of unresponsive flows occur, the correlation between these metrics changes significantly and is much less obvious in general. Therefore, in the present paper, we have decided to keep the very same set of metrics altogether even though they are not always linearly independent.

3 Dealing With Nuisance Parameters

Generally speaking, the problem we address in the paper is the detection of a signal (the impact of unresponsive traffic) in the presence of so-called nuisance parameters. Indeed, legitimate traffic constitutes a complex "non-anomalous" environment: the legitimate traffic is not constant: it can change depending on the users' behaviours. The legitimate traffic is generally unknown and it is not straightforward to model. In the present paper, we adopt a general statistical model: the legitimate traffic is assumed to be drawn from the following Gaussian distribution:

$$\mathbf{x}_t \sim \mathcal{N}(\boldsymbol{\mu}_t, \boldsymbol{\Sigma}_t), \quad \text{under legitimate traffic,} \tag{1}$$

where $\mathcal{N}(\cdot)$ represents the Gaussian, or normal, distribution, $\mathbf{x}_t$ represents the metrics at time t whose expected value is denoted $\boldsymbol{\mu}_t$ and with covariance matrix $\boldsymbol{\Sigma}_t$.

On the opposite, when an attack on low-latency services is launched, the traffic is modelled by a sum of the legitimate traffic and the impact of the attack:

$$\mathbf{x}_t \sim \mathcal{N}(\boldsymbol{\mu}_t + \mathbf{a}_t, \boldsymbol{\Sigma}_t), \quad \text{under attack.} \tag{2}$$

In other words, it is assumed that the attack impacts the expectation of the observation and not much the covariance matrix, which remains almost unchanged. In addition, the anomaly $\mathbf{a}_t$ is assumed constant ; more precisely, the impact of the attack can vary in intensity but its "footprint" remains the same, hence $\mathbf{a}_t = \eta_t \mathbf{a}$ with $\mathbf{a}$ the constant "footprint" and η model the "intensity" of the attack.

Usually, the first step when dealing with correlated observations is to apply the so-called whitening transformation with the matrix $\boldsymbol{\Sigma}_t^{-1/2}$ satisfying $\boldsymbol{\Sigma}_t^{-1/2^\top}\boldsymbol{\Sigma}_t^{-1/2} = \boldsymbol{\Sigma}_t$. Usually, $\boldsymbol{\Sigma}_t^{-1/2}$ is obtained as the Chelesky decomposition of the covariance matrix $\boldsymbol{\Sigma}_t$. Indeed, it is straightforward from (1)–(2) that:

$$\mathbf{x}_t' = \boldsymbol{\Sigma}_t^{-1/2^\top}\mathbf{x}_t' \sim \mathcal{N}(\boldsymbol{\mu}_t', \mathbf{I}), \quad \text{under legitimate traffic,} \tag{3}$$

With $\boldsymbol{\mu}_t' = \boldsymbol{\Sigma}_t^{-1/2^\top}\boldsymbol{\mu}_t$.

It is obvious from Eqs. (1)–(3) that the legitimate traffic $\boldsymbol{\mu}_t$ represents a nuisance parameter in the sense that it has no interest for the detection of the signal of interest, namely the attack footprint $\mathbf{a}$, but it shall be carefully taken into account, as it may obfuscate the attack it is aimed at detecting.

In the present paper, we adopt a linear model of the legitimate traffic:

$$\boldsymbol{\mu}_t = \mathbf{H}_t\theta_t, \tag{4}$$

where the column of the matrix $\mathbf{H}_t$ spans the subspace of the legitimate traffic at time t.

The main advantage of the linear model (4) is that it easily allows removing the nuisance parameters $\boldsymbol{\mu}_t = \mathbf{H}_t\theta_t$. Indeed, the rejection of the nuisance parameter can be carried out by simply projecting the vector of network traffic measurements $\mathbf{x}_t$ onto the orthogonal complement $R(\mathbf{H}_t)^\perp$ of the legitimate traffic, which is the subspace spanned by the column matrix $\mathbf{H}_t$. This projection can be defined by the projector of the Generalized Least-Square (GLS) as follows:

$$\mathbf{P}_{\mathbf{H}_t}^\perp = \mathbf{I}_p - \mathbf{H}_t\left(\mathbf{H}_t^\top\boldsymbol{\Sigma}_t^{-1}\mathbf{H}_t\right)^{-1}\mathbf{H}_t^\top\boldsymbol{\Sigma}_t^{-1}. \tag{5}$$

Obviously, the second term of matrix $\mathbf{P}_{\mathbf{H}_t}^\perp$ corresponds to the least square estimation of the nuisance parameter $\mathbf{H}_t\theta_t$ under the assumption that the covariance matrix $\boldsymbol{\Sigma}_t$ is proportional to the identity matrix. Hence, the projector $\mathbf{P}_{\mathbf{H}_t}^\perp$ corresponds to the subtraction of this estimation of legitimate traffic from the observation.

Alternatively, the rejection of the nuisance parameter can be achieved with the orthonormal matrix $\mathbf{W}_t = (w_1, \ldots, wp - q)$, where w_i are the eigenvectors of the projection matrix $\mathbf{P}_{\mathbf{H}_t}^\perp$ corresponding to eigenvalues 1.

The matrix $\mathbf{W}$ verifies the following properties:

$$\mathbf{WH} = \mathbf{0}\ , \quad \mathbf{W}^\top\mathbf{W} = \mathbf{PH}^\perp\ , \quad \mathbf{WW}^\top = \mathbf{I}p - q. \tag{6}$$

The rejection of a linear nuisance parameter can be simply carried out as

$$\mathbf{W}_t^\top\mathbf{x}_t. \tag{7}$$

It immediately follows from Eqs. (1), (2) and (7) that:

$$\mathbf{W}_t^\top \mathbf{x}_t' \sim \mathcal{N}(\mathbf{0}, \mathbf{I}p - q), \quad \text{under legitimate traffic,} \tag{8}$$

and:

$$\mathbf{W}_t^\top \mathbf{x}_t' \sim \mathcal{N}(\mathbf{a}^\perp, \mathbf{I}p - q), \quad \text{under attack,} \tag{9}$$

where $\mathbf{a}^\perp = \mathbf{W}_t^\top \mathbf{a}$ represents the impact of the attack on low-latency services it is aimed at detecting.

4 Proposed Autoadaptive Linear Model for Representing Network Traffic

The model presented in the previous Sect. 3 is very general in statistical decision theory [12–14] and has been widely used in image processing [15–20] and computer network traffic modelling [21–26]. However, it is not accurate enough for the targeted application of detection of attacks on low-latency services to represent the network traffic used in the present work.

To design a model that is accurate enough to represent the legitimate network traffic and, yet, to preserve the simplicity of the linearity, we propose an original autoadaptive model [22,27–29]. The principle of this model is depicted in Fig. 1. First the measurements are gathered and the set of L last measurements $\mathbf{X}_t = \mathbf{x}_{t-L+1}, \mathbf{x}_{t-L+2}, \ldots, \mathbf{x}_t$ are analysed jointly. These samples are modelled altogether using a linear parametric model:

$$\mathbf{X}_t \sim \mathcal{N}(\mathbf{H}_t \theta_t, \mathbf{\Sigma}_t), \quad \text{under legitimate traffic,} \tag{10}$$

The $M = kL$ previous samples are used to (1) estimate the covariance matrix $\mathbf{\Sigma}_t$ and to build the linear model $\mathbf{H}_t$ adaptively. To this end we perform a principal component analysis (PCA) on the k windows of size L using the $M = kL$ previous observations $x_{t-2L-M+1}, \ldots, x_{t-2L},$. Let us denote $\mathbf{p}_1, \ldots, \mathbf{p}_L$ the principal components sorted by the associated eigenvalues. The linear model is built with the n principal components $\mathbf{H}_t = (\mathbf{p}_1, \ldots, \mathbf{p}_n)$.

Note in our case, we tested several configurations, and we kept the one that gave the best results: $L = 8$ for a total of $p = 64$ metrics analysed jointly with a number of components $n = 3$.

Also, note that, as depicted in Fig. 1 some between the "detection window" $\mathbf{x}_{t-L+1}, \ldots, \mathbf{x}_t$ and the "parameters estimation window" $x_{t-2L-M+1}, \ldots, x_{t-2L},$ are not used. Indeed, the idea of the proposed autoadaptive linear model we proposed is to carry out the estimation and the rejection with observations under legitimate traffic, in other words, that are free from the impact of a potential attack on low-latency services. A certain detection delay may be needed to ensure that the attack is detected; hence the need for a certain distance between the parameter estimation window and the observation used for the detection.

With this idea of using samples without the impact of a potential attack, the last originality of the proposed autoadaptive linear model is to fix the parameter

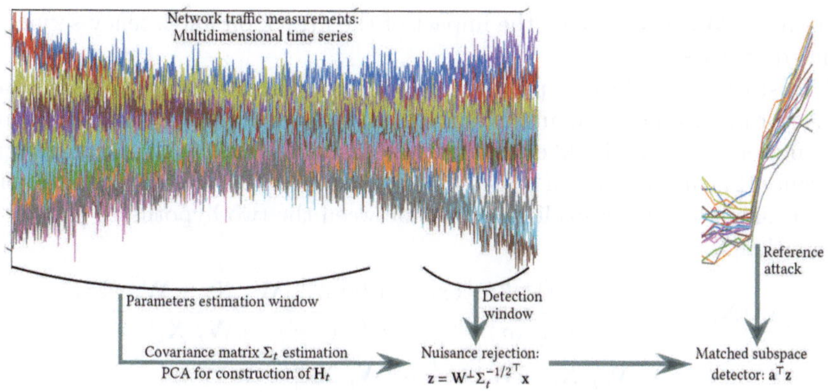

Fig. 1. Illustration of the architecture of the proposed model. The network traffic measurements, represented as a multivariate time series, are firstly input into the autoencoder, whose goal is to learn the underlying model and representation of legitimate traffic. Then the residuals are considered by subtracting the output of this autoencoder and the result is fed to a multi-headed self-attention module before classification.

estimation window when an attack is detected. While the two windows are "sliding" with the observation under legitimate traffic, we keep on using the same samples for parameter estimation as long as an attack is detected under the "detection window" to ensure that the estimation is, as much as possible, free from the impact of a potential attack.

Last but not least, note that we assume that noise corrupting the metrics is not correlated in time; hence the statistical independence between x_{t-1} and x_t. This allows us to reduce the estimation of the covariance matrix $\mathbf{\Sigma}_t$ by the covariance between metrics.

5 Methodology for Detection of Attacks on Low-Latency Services

Using the original autoadaptive model described in Sect. 4 we can eventually describe the observed metrics of the computer networks as:

$$\begin{cases} \text{under } \mathcal{H}_0 : \mathbf{X}_t \sim \mathcal{N}(\mathbf{H}_t\theta_t, \mathbf{\Sigma}_t), \\ \text{under } \mathcal{H}_1 : \mathbf{X}_t \sim \mathcal{N}(\mathbf{H}_t\theta_t + \mathbf{a}, \mathbf{\Sigma}_t). \end{cases} \tag{11}$$

As explained in Sect. 3 after whitening transformation and rejection of the legitimate traffic $\mathbf{H}_t\theta_t$ the statistical detection problem(11) can be rewritten, under the principle of invariance, see [30][Sect. 4], as:

$$\begin{cases} \text{under } \mathcal{H}_0 : \mathbf{W}_t^\top \mathbf{X}_t \sim \mathcal{N}(\mathbf{0}, \mathbf{I}), \\ \text{under } \mathcal{H}_1 : \mathbf{W}_t^\top \mathbf{X}_t \sim \mathcal{N}(\mathbf{a}^\perp, \mathbf{0}). \end{cases} \tag{12}$$

where $\mathbf{a}^{\perp} = \mathbf{W}_t^{\top}\mathbf{a}$ represents the impact of the attack on low-latency services it is aimed at detecting.

The statistical detection problem (12) is a test between two simple hypotheses. According to the Neyman-Pearson Lemma [30, Theorem 3.2.1], the optimal, most powerful test is the likelihood ratio test. Because the statistical detection problem (12) can be reduced to detecting a known signal $\mathbf{a}^{\perp}$ in Gaussian noise, we can calculate the log-likelihood ratio between the two hypotheses (12) for an observation $\mathbf{X}_t$ as:

$$\Lambda(\mathbf{X}_t) = \log \frac{(2\sqrt{\pi}\sigma)^{-(p-q)} \exp\left(-(2\sigma^2)^{-1}\|\mathbf{W}_t^{\top}\mathbf{X}_t - \mathbf{W}_t^{\top}\mathbf{a}\|^2\right)}{(2\sqrt{\pi}\sigma)^{-(p-q)} \exp\left(-(2\sigma^2)^{-1}\|\mathbf{W}_t^{\top}\mathbf{X}_t\|^2\right)} \tag{13}$$

$$= -\|\mathbf{W}_t^{\top}\mathbf{X}_t - \mathbf{W}_t^{\top}\mathbf{a}\|^2 + \|\mathbf{W}_t^{\top}\mathbf{X}_t\|^2 \tag{14}$$

where $\|\mathbf{x}\|_2^2 = \mathbf{x}^{\top}\mathbf{x}$ is the Euclidean norm, $p = 64$ is the dimension of $\mathbf{X}_t$ and $n = 3$ is the number of principal components used to design the matrix $\mathbf{H}_t$.

Obviously, the log-likelihood ratio established in the Eq. (14) depends on the observation $\mathbf{X}_t$ only through the term:

$$\Lambda^{\star}(\mathbf{X}_t) = \left(\mathbf{W}_t^{\top}\mathbf{X}_t\right)^{\top}\mathbf{W}_t^{\top}\mathbf{a} = \mathbf{X}_t^{\top}\mathbf{W}_t\mathbf{W}_t^{\top}\mathbf{a} = \mathbf{X}_t^{\top}\mathbf{P}_{\mathbf{H}}^{\perp}\mathbf{a}, \tag{15}$$

which correspond to the well-known matched subspace detectors [13] after rejection of the non-anomalous background via $\mathbf{W}_t^{\top}$.

In practice the covariance matrix $\mathbf{\Sigma}_t^{-1/2^{\top}}$ and the linear model $\mathbf{H}_t$ are estimated from the previous observations, as explained in the Sect. 4.

Similarly, since the network traffic is widely varying the decision statistics $\Lambda^{\star}(\mathbf{X}_t)$ do not always have a zero mean under legitimate traffic. Therefore, the "parameter estimation window" is used to estimate the mean of the decision statistic $\Lambda_0^{\star}$ which is subtracted from the LR calculated over the detection window:

$$\Lambda^{\star}(\mathbf{X}_t) = \mathbf{X}_t^{\top}\mathbf{P}_{\mathbf{H}}^{\perp}\mathbf{a} - \Lambda_0^{\star}. \tag{16}$$

The statistical test is finally defined formally by:

$$\delta(\mathbf{X}_t) = \begin{cases} \mathcal{H}_0 & \text{if } \Lambda^{\star}(\mathbf{X}_t) = \mathbf{X}_t^{\top}\mathbf{P}_{\mathbf{H}}^{\perp}\mathbf{a} - \Lambda_0^{\star} < \tau \\ \mathcal{H}_1 & \text{if } \Lambda^{\star}(\mathbf{X}_t) = \mathbf{X}_t^{\top}\mathbf{P}_{\mathbf{H}}^{\perp}\mathbf{a} - \Lambda_0^{\star} \geq \tau \end{cases} \tag{17}$$

where τ is the decision threshold.

The detection methodology is summarized in Algorithm 1 which clearly explains the role of the estimation window, the detection window and how the statistical test (14)–(16) is calculated using real network traffic metrics.

6 Numerical Experimentation and Results

6.1 Experimental Setup

We built a testbed, shown in Fig. 2, with virtual machines for clients and servers, and a bare-metal switch that uses P4 language[1] to implement L4S AQM, routing, and forwarding functions. This allows us to monitor flows at a packet-level

[1] P4.org.

Algorithm 1. Low-lantency attack detection using the autoadaptive model

Require: L ▷ Detection window size
Require: $M = kL$ ▷ Estimation window size
Require: $\mathbf{X}_t = (x_{t-L+1}, \dots, x_t)$ ▷ Detection window
Require: $\mathbf{X}_0 = (x_{n-M+1}, \dots, x_n) = (\mathbf{X}_1, \dots, \mathbf{X}_k)$ ▷ Estimation window
  Using the estimation window $\mathbf{X}_0$:
  $\mathbf{H}_t \leftarrow PCA(\mathbf{X}_1, \dots, \mathbf{X}_k)$
  $\mathbf{\Sigma}_t \leftarrow Covariance(\mathbf{X}_0)$
  $\mathbf{P}_{\mathbf{H}_t}^{\perp} \leftarrow \mathbf{I}_p - \mathbf{H}_t \left(\mathbf{H}_t^{\top} \mathbf{\Sigma}_t^{-1} \mathbf{H}_t\right)^{-1} \mathbf{H}_t^{\top} \mathbf{\Sigma}_t^{-1}$
  $\Lambda_0^{\star} \leftarrow Mean(\mathbf{X}_1^{\perp} \mathbf{P}_{\mathbf{H}_t}^{\perp} \mathbf{a}, \dots, \mathbf{X}_k^{\perp} \mathbf{P}_{\mathbf{H}_t}^{\perp} \mathbf{a})$
  Using the detection window $\mathbf{X}_t$:
  $\Lambda^{\star}(\mathbf{X}_t) \leftarrow \mathbf{X}_t^{\perp} \mathbf{P}_{\mathbf{H}_t}^{\perp} \mathbf{a} - \Lambda_0^{\star}$
  if $\Lambda^{\star}(\mathbf{X}_t) \leq \tau$ **then**
    $n = t - M - L$ ▷ The estimation window is moved to the previous metrics
  else if $\Lambda^{\star}(\mathbf{X}_t) > \tau$ **then**
    $\mathbf{X}_0 \leftarrow \mathbf{X}_0$ ▷ Under attack $\mathbf{X}_0$ remains unchanged
  end if

granularity using In-band Network Telemetry (INT). The router runs the P4 implementation of DualPI2, as specified by the IETF. Users can choose what data to collect, and the INT system sends reports to a collector. More details on the testbed can be found in this article [31].

Our testbed; depicted in the Fig. 2, has several legitimate clients and servers, some using low-latency services and classic flows. We measure everything from the central L4S-capable router. We vary the number of legitimate clients and the attacker/regular ratio from 1:1 to 1:10. We also vary the attack power from the minimal attack power, behaving like a legitimate flow by reducing its rate according to the Prague requirements, to the maximum attack power, by not reducing, at all, its rate regardless of the number of ECN signals. This means we can create a range of attack scenarios, from weak attacks with one attack flow hidden in ten legitimate flows, to strong attacks where several attackers compete with one legitimate flow for bandwidth. This setting lets us test different attack types, from subtle to obvious, and see how they affect the network."

We conducted about 200 experiments, each lasting 90 s. The experiment starts with 30 s of normal traffic, followed by 30 s of attack traffic, and ends with 30 s of normal traffic again. This design allows us to capture the attack's early stages when its effects are not yet fully visible, making it harder to detect. The return to normal traffic also helps us study the aftermath of an attack, including the time it takes for the router to recover. We collected traffic data during these transition periods, which can be slightly different. We note that this experimental approach generated a lot more normal traffic data than attack data. Some experiments had no attack at all, serving as a reference point. Our large-scale experiment resulted in a total of 714, 967 samples of attack traffic and 3, 901, 982 samples of normal traffic. This means that about 15% of the samples were

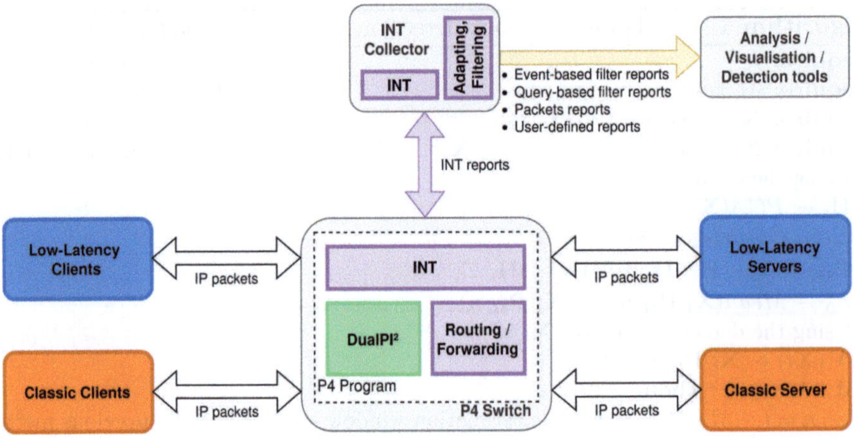

Fig. 2. Illustration of our experimental testbed ; note that the number of legitimate users can vary as well as the traffic and the available achievable transmission rate of the router.

under attack, while about 85% were normal traffic, as reported in our previous work [31].

6.2 Experimental Detection Results

The very first interesting results concern the relevance of the proposed methodology. We especially want to highlight the importance of the two windows procedure where the first window is used for parameter estimation and the second window is the one over which the detection is carried out. To this end the Fig. 3 contrasts the detection results for one single experimentation. Note that the results focus on a small period before and after the attack starts. The red curve shows the detection statistics $\Lambda^\star$ as given in the Eq. (15) when not subtracting the mean Log-LR $\lambda_0^\star$ over the estimation window. Clearly, the impact of the attack is easily detectable. Note, however, the non-zero mean before the kick-off of the attack. The average value of the Log-LR changes for every experimentation, which makes it very difficult to set a threshold τ above which it is assumed that an attack is happening. Indeed, for some experimentation, the mean of the Log-LR $\Lambda^\star$ is large before the attack, creating many false positives. On the opposite, for some other experiments, the mean of the log-LR is very small and the impact of the attack to small preventing the detection of the attack.

The blue curve shows the values of the log-LR $\Lambda^\star$ when the estimation window keeps moving even when an attack is detected. Obviously, the attack is clearly detected but its impact fades away very quickly because the estimation windows start containing observations after the attack star. Therefore the adaptive model quickly takes into account the new observations and the impact of the attack is slowly incorporated into the linear model $\mathbf{H}_t$. One can note that is such a case if the attack is not detected as soon as it starts, it is very unlikely to detect it after a small delay.

The green curve shows the value of the proposed overall methodology for calculating the log-LR $\Lambda^\star$ taking into account the average value of the decision statistics over the estimation window and the keeping the same estimation window when an attack is detected. Obviously, the proposed methodology, using the estimation windows to compute the log-LR $\lambda_0^\star$ as in Eq. (16) and as described in Algorithm 1, is very relevant to obtain a standard decision statistic that is almost always centered around zero under legitimate traffic only and preserves the detection possibility even after a few seconds of the attack kick-off.

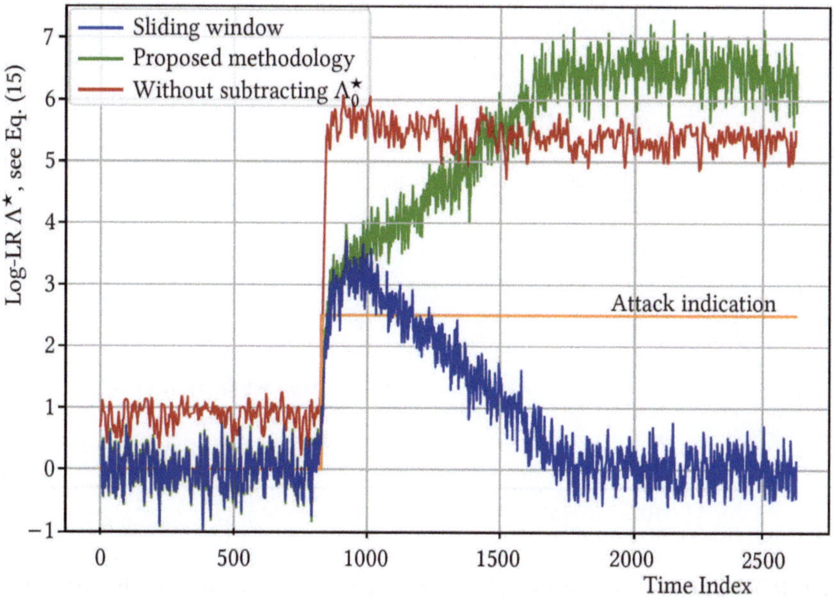

Fig. 3. Comparison of the proposed method with ablated versions ; the green curve shows the log-LR $\Lambda_0^s tar$ computed using the proposed methodology ; in red, the methodology is always applied using the last observation for the estimation window even under detection of the attack ; in blue, the proposed methodology is used without subtracting the mean value of the log-LR $\mathbf{a}_0^\top$ estimated over the estimation window.

The most important results about the detection performance of the proposed original methodology are presented in Fig. 4. This figure presents the detection performance, using three different window sizes, as ROC (Receiver Operating Characteristics) curves. These curves present the detection accuracy, measured as the true positive rate, also referred to as the power function, the test sensitivity or the recall, as a function of the false alarm rate. Note that for readability, the Fig. 4 uses a semi-logarithmic scale for the false alarm rate ; this allows better emphasizing the very high detection power achieved for a false alarm rate as small as 0.03 which would be clearly not very visible using a linear scale.

Interestingly, the Fig. 4 shows that for a vast majority of the cases the attack on low-latency services can be detected. It also shows that there are small, but not negligible, fraction of cases in which detection is not significantly better as random guessing. This seems to point out that about 3% of network traffic observations under attack are similar to those without attack ; this often occurs at the end of the attack and is in part explainable by issues regarding the labelisation of the data, as sometimes the attack does not stop exactly after 30 seconds.

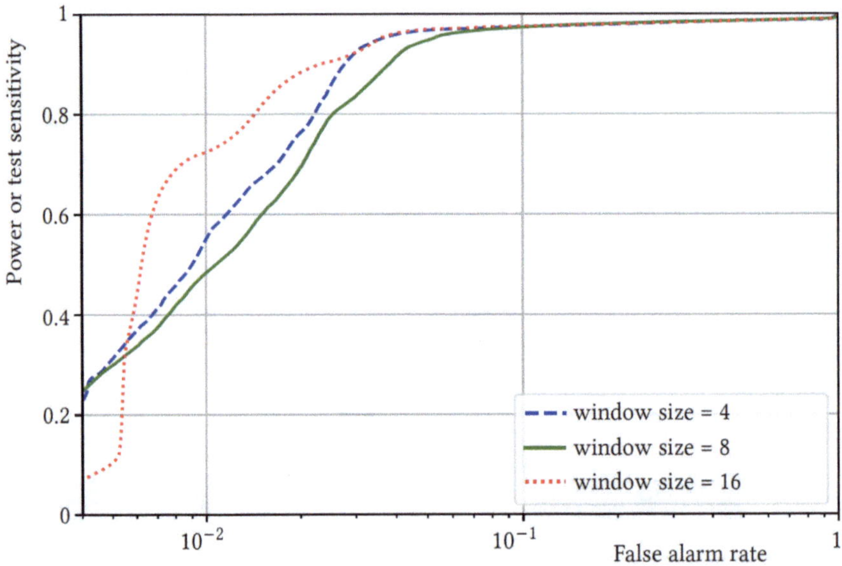

Fig. 4. ROC Curves of the proposed detector methodology applied with different window sizes.

To show the limit of the proposed methodology, the Fig. 5 shows the false alarm rate as a function of the detection threshold τ. Note that, interestingly, Fig. 5 shows, in cyan, a comparison with the theoretical false alarm rate one would obtain if the observation follows exactly the Gaussian distribution model presented in the Eqs. (11)–(12). This result clearly emphasizes the limit of the proposed methodology, as the theoretical false alarm rate differs significantly from observations for a false alarm rate smaller than 10%. This shows that the assumed statistical linear model is not accurate enough to allow warranting a false alarm rate setting the decision threshold τ according to the theoretical model. This is due to, in part, to the fact that the observation does not always follow Gaussian distribution and, in part, to the limit of the estimated autoadaptive linear model $\mathbf{H}_t$ to represent the behaviour of the network traffic metrics.

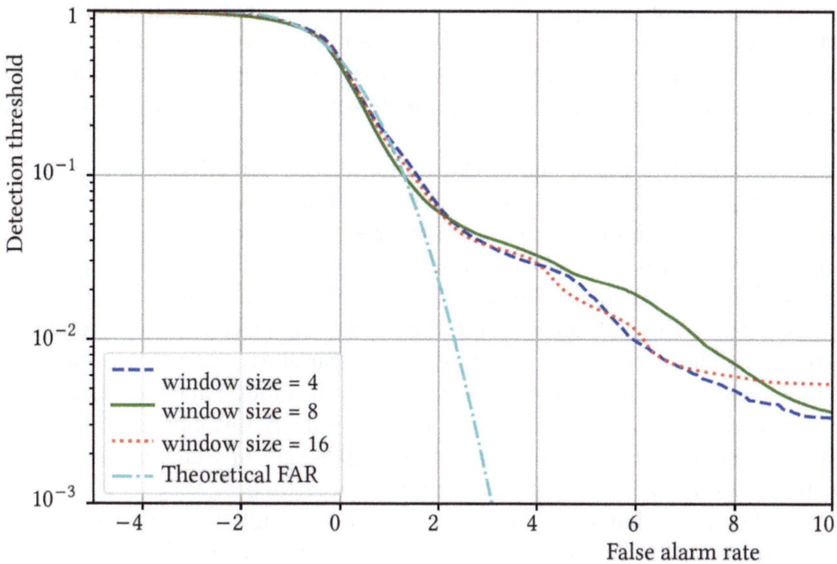

Fig. 5. ROC Curves of the proposed simple model with ablation versions. Note that the x-axis representing the false positive rate is plotted using a logarithmic scale.

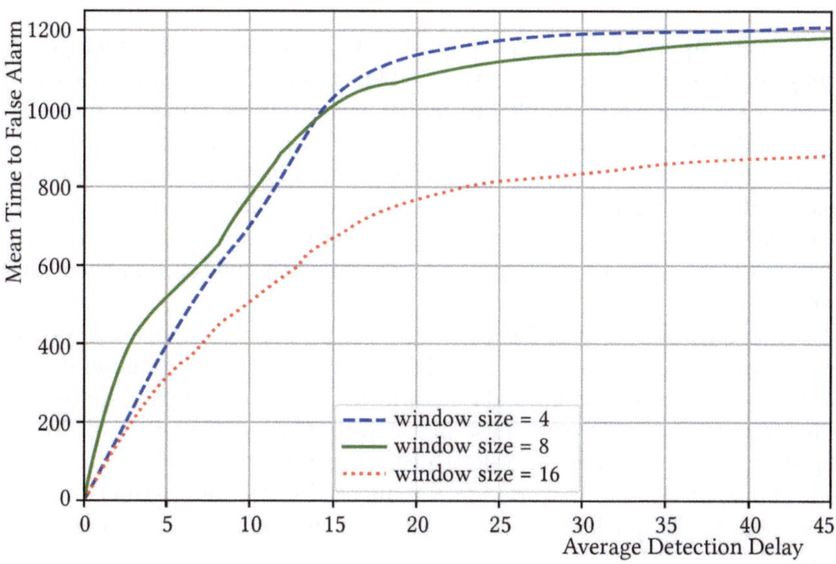

Fig. 6. ROC Curves of the proposed simple model with ablation versions. Note that the x-axis representing the false positive rate is plotted using a logarithmic scale.

The last interesting results we wanted to show concern the detection delay. Even though the present paper does not propose a specific sequential detection method, such as the well-known CUSUM (cumulative sum), it is interesting to measure, for a given detection threshold τ, the average time before a false alarm is raised and the average delay for detecting the beginning of the attack on low-latency services.

To this end, the Fig. 6 shows the result on the detection delay as follows: the average time to false alarm is plotted as a function of the average detection delay. Note that the delays, here, are represented in terms of the number of samples. Ideally, one wishes to minimize the detection delay while having a very large average time to false alarm. Therefore, best results are obtained on the top left corner. Figure 6 shows that the detection delay is relatively small as compared to the average time to false alarm. Similarly, this figure shows that the best results are obtained for a window size of about 8 measurements of the metrics ; a smaller window reduces the detection delay but increases the number of false alarm hence the lower average time to false alarm while, on the opposite, a higher window size delays the detection of the attack.

Note, however, that with the experimentation conducted for this paper is it difficult to measure the average time to false alarm higher than 30 s, while in real life conditions such an average period between false alarms is far too small. Also, note that better results would have been obtained using a proper sequential detection method, which is not the scope of the present paper.

7 Conclusion and Future Works

The present paper addresses the problem of the detection of attacks on low-latency services in computer networks. This problem is difficult because the legitimate traffic is very complex to model accurately in a general manner. To cope with this issue, we proposed a simple yet efficient autoadaptive linear model for representing the legitimate traffic rather accurately. Using this linear model of the network traffic metrics, we proposed a detection methodology using two sliding windows, the first being used for estimating the parameters of the autoadaptive model and the second is the one over which the presence of the attack is detected.

Over a large set of real-life experimentation, we show that the proposed original methodology, despite its simplicity, achieves rather very good detection performance. However, we have also shown the limitations of the proposed approach for guaranteeing a very false alarm rate. Our next works will focus on using deep learning models in order to compare the performance with the proposed approach as well as to design a sequential detection method to address the problem of the quickest detection under a constraint on the average time to alarm.

Acknowledgments. This work was partially supported by the French ANR MOSAICO project No. ANR-19-CE25-0012.

References

1. Albisser, O., De Schepper, K., Briscoe, B., Tilmans, O., Steen, H.: Dualpi2-low latency, low loss and scalable throughput (l4s) AQM. Proc. of Netdev (2019)
2. De Schepper, K., Bagnulo, M., White, G.: RFC 9330: Low Latency, Low Loss, and Scalable Throughput (L4S) Internet Service: Architecture. RFC Editor (2023)
3. Mathieu, B., Tuffin, S.: Evaluating the l4s architecture in cellular networks with a programmable switch. In: 2021 IEEE Symposium on Computers and Communications (ISCC), pp. 1–6 (2021). https://doi.org/10.1109/ISCC53001.2021.9631539
4. BoruOljira, D., Grinnemo, K.-J., Brunstrom, A., Taheri, J.: Validating the sharing behavior and latency characteristics of the l4s architecture. ACM SIGCOMM Comput. Commun. Rev. **50**(2), 37–44 (2020)
5. Hutchings, A., Clayton, R.: Exploring the provision of online booter services. Deviant Behav. **37**(10), 1163–1178 (2016)
6. Letourneau, M., N'Djore, K.B., Doyen, G., Mathieu, B., Cogranne, R., Nguyen, H.N.: Assessing the threats targeting low latency traffic: the case of l4s. In: 2021 17th International Conference on Network and Service Management (CNSM), pp. 544–550 (2021). IEEE
7. Letourneau, M., Doyen, G., Cogranne, R., Mathieu, B.: A comprehensive characterization of threats targeting low-latency services: the case of l4s. J. Netw. Syst. Manage. **31**(1), 19 (2023)
8. Sarpkaya, F.B., Srivastava, A., Fund, F., Panwar, S.: To switch or not to switch to TCP prague? Incentives for adoption in a partial l4s deployment. In: Proceedings of the 2024 Applied Networking Research Workshop. ANRW '24, pp. 45–52. Association for Computing Machinery, New York, NY, USA (2024). https://doi.org/10.1145/3673422.3674896
9. Nádas, S., Gombos, G., Fejes, F., Laki, S.: A congestion control independent l4s scheduler. In: Proceedings of the 2020 Applied Networking Research Workshop. ANRW '20, pp. 45–51. Association for Computing Machinery, New York, NY, USA (2020). https://doi.org/10.1145/3404868.3406669
10. De Schepper, K., Bondarenko, O., Tsang, I.-J., Briscoe, B.: PI2: a linearized AQM for both classic and scalable TCP. In: Proceedings of the 12th International on Conference on Emerging Networking EXperiments and Technologies. CoNEXT '16, pp. 105–119. Association for Computing Machinery, New York, NY, USA (2016). https://doi.org/10.1145/2999572.2999578
11. Schepper, K.D., Briscoe, B., White, G.: DualQ Coupled AQMs for Low Latency, Low Loss and Scalable Throughput (L4S) Internet Service: Architecture
12. Frank, P.M.: Fault diagnosis in dynamic systems using analytical and knowledge-based redundancy: a survey and some new results. Automatica **26**(3), 459–474 (1990)
13. Scharf, L.L., Friedlander, B.: Matched subspace detectors. IEEE Trans. Signal Process. **42**(8), 2146–2157 (1994). https://doi.org/10.1109/78.301849
14. Fouladirad, M., Nikiforov, I.: Optimal statistical fault detection with nuisance parameters. Automatica **41**(7), 1157–1171 (2005)
15. Cogranne, R., Retraint, F.: Statistical detection of defects in radiographic images using an adaptive parametric model. Signal Process. **96**, 173–189 (2014). https://doi.org/10.1016/j.sigpro.2013.09.016
16. Cogranne, R., Retraint, F.: Statistical detection of defects in radiographic images using an adaptive parametric model. Signal Process. **96**, 173–189 (2014)

17. Sedighi, V., Fridrich, J., Cogranne, R.: Content-adaptive pentary steganography using the multivariate generalized gaussian cover model. In: Media Watermarking, Security, and Forensics 2015, vol. 9409, pp. 144–156 (2015). SPIE
18. Sedighi, V., Cogranne, R., Fridrich, J.: Content-adaptive steganography by minimizing statistical detectability. IEEE Trans. Inf. Forensics Secur. **11**(2), 221–234 (2015)
19. Cogranne, R., Giboulot, Q., Bas, P.: Steganography by minimizing statistical detectability: the cases of jpeg and color images. In: Proceedings of the 2020 ACM Workshop on Information Hiding and Multimedia Security. IH & MMSec '20, pp. 161–167. Association for Computing Machinery, New York, NY, USA (2020). https://doi.org/10.1145/3369412.3395075
20. Cogranne, R., Giboulot, Q., Bas, P.: Efficient steganography in JPEG images by minimizing performance of optimal detector. IEEE Trans. Inf. Foren. Secur. **17**, 1328–1343 (2021)
21. Basu, S., Mukherjee, A., Klivansky, S.: Time series models for internet traffic. In: Proceedings of IEEE INFOCOM'96. Conference on Computer Communications, vol. 2, pp. 611–620 (1996). IEEE
22. Yin, H., Lin, C., Sebastien, B., Li, B., Min, G.: Network traffic prediction based on a new time series model. Int. J. Commun Syst **18**(8), 711–729 (2005)
23. Nguyen, T., Cogranne, R., Doyen, G.: An optimal statistical test for robust detection against interest flooding attacks in CCN. In: 2015 IFIP/IEEE International Symposium on Integrated Network Management (IM), pp. 252–260 (2015). IEEE
24. Nguyen, T., et al.: Reliable detection of interest flooding attack in real deployment of named data networking. IEEE Trans. Inf. Forensics Secur. **14**(9), 2470–2489 (2019). https://doi.org/10.1109/TIFS.2019.2899247
25. Lian, L.: Network traffic prediction model based on linear and nonlinear model combination. ETRI J. **46**(3), 461–472 (2024). https://doi.org/10.4218/etrij.2023-0136
26. Azab, A., Khasawneh, M., Alrabaee, S., Choo, K.-K.R., Sarsour, M.: Network traffic classification: techniques, datasets, and challenges. Digit. Commun. Netw. **10**(3), 676–692 (2024)
27. Cogranne, R., Doyen, G., Ghadban, N., Hammi, B.: Detecting Botclouds at large scale: a decentralized and robust detection method for multi-tenant virtualized environments. IEEE Trans. Netw. Serv. Manage. **15**(1), 68–82 (2018). https://doi.org/10.1109/TNSM.2017.2785628
28. Tout, K., Cogranne, R., Retraint, F.: Statistical decision methods in the presence of linear nuisance parameters and despite imaging system heteroscedastic noise: application to wheel surface inspection. Signal Process. **144**, 430–443 (2018). https://doi.org/10.1016/j.sigpro.2017.10.030
29. Nguyen, T., et al.: A security monitoring plane for named data networking deployment. IEEE Commun. Mag. **56**(11), 88–94 (2018)
30. Lehmann, E.L., Romano, J.P.: Testing Statistical Hypotheses, 3rd Edn. Springer, (2005)
31. Nguyen, H.N., Mathieu, B., Letourneau, M., Doyen, G., Tuffin, S., de Oca, E.M.: A comprehensive P4-based monitoring framework for l4s leveraging in-band network telemetry. In: NOMS 2023-2023 IEEE/IFIP Network Operations and Management Symposium, pp. 1–6 (2023). IEEE

Automated Food Image Labeling for E-Commerce Websites: Combining Content-Based Image Retrieval and Majority Labeling

Khang Nguyen Hoang(✉), Huynh Vu Nhu Nguyen, and Hoang Ngoc Tran

FPT University, Cantho, Vietnam
{khangnhce171197,huynhnvnce170550}@fpt.edu.vn

Abstract. E-commerce platforms frequently update food images to improve user experience, but manual labeling is costly and time-consuming. This study presents an automated labeling system combining Content-Based Image Retrieval (CBIR) and Vision Transformer (ViT) for feature extraction. ViT-Large (Patch 16) achieved superior performance with 0.8334 mAP and 0.6910 Precision, outperforming CNNs like EfficientNet and ResNet. A vector database enables fast retrieval, and a temporary labeling mechanism assigns high-confidence labels from top-k similar images. These labels are later used to retrain ViT models, ensuring continuous improvement. The system reduces labeling costs, maintains data quality, and enhances update efficiency for food-related e-commerce applications.

Keywords: Vision Transformers (ViTs) · Convolutional Neural Networks (CNNs) · 30VNFood dataset · Vietnamese cuisine · Fine-grained image classification · Deep learning

1 Introduction

Food images are vital in e-commerce, especially in the food industry, as they influence customer engagement and purchasing decisions. High-quality, regularly updated images enhance both product understanding and the shopping experience. However, manually labeling food images is time-consuming, costly, and error-prone. As e-commerce grows, the volume of new image data makes traditional manual labeling unsustainable. (Xie et al., 2020; Bilen & Vedaldi, 2016) [1, 2].

These challenges highlight the urgent need for a more efficient solution that can automate the process of updating and labeling food images. To address this issue, this study proposes an automated system that leverages advanced image retrieval techniques and a temporary labeling mechanism to classify and label food images effectively. By automating this process, the system aims to reduce labor costs, improve data update efficiency, and enhance the user experience on food e-commerce websites. This approach not only tackles the inherent limitations of manual labeling but also ensures that product image data remains current and reliable. Furthermore, the system must support

A. Ito (Ed.): ICMSSP 2025, LNNS 1637, pp. 59–66, 2025.
https://doi.org/10.1007/978-3-032-05994-9_6

content-based search functionality within the database, enabling users to find similar food items based on image content rather than relying solely on textual keywords. This dual capability enhances the utility of the system by providing seamless and accurate search experiences for end users (Qiu et al., 2019; Matsuda & Yanai, 2012) [3, 4].

The main contributions of this study include the development of an automated system for updating food image data on e-commerce websites, addressing the challenges of manual image labeling. The system's effectiveness is evaluated using standard metrics in image retrieval, such as precision and mAP on Top-K. Additionally, the study provides a comparative analysis of ViT against other models, examines the impact of temporary labeling on system performance, and proposes enhancements to improve image quality and system efficiency (Bossard et al., 2014; Thai et al., 2017) [5, 6].

2 Related Work

2.1 Image Retrieval

Image retrieval focuses on searching for images based on their visual content rather than metadata or textual descriptions. Traditional techniques for image feature extraction include SIFT, SURF, and HOG, which rely on handcrafted features (Dalal & Triggs, 2005) [7]. However, with the rise of deep learning, feature extraction has significantly improved, leveraging convolutional neural networks (CNNs) like ResNet and EfficientNet, and more recently, Vision Transformers (ViTs). These modern architectures offer hierarchical and abstract representations of image content, making them highly effective for retrieval tasks (Dosovitskiy et al., 2021; Tan & Le, 2019) [8, 9].

The similarity between feature vectors is typically measured using metrics such as Euclidean distance, cosine similarity, or Hamming distance. Evaluating the effectiveness of image retrieval systems commonly involves metrics like Precision for top-k retrieval, Recall for top-k retrieval, F1-score for top-k retrieval, and mean Average Precision (mAP). The concept of "top-k retrieval" refers to retrieving the top k most relevant images for a given query, where k represents the number of results returned. For example, in Precision for Top-30 retrieval, the system's ability to correctly retrieve 30 relevant images is assessed. These metrics provide a clear understanding of how well the system performs in delivering accurate and relevant results to the user (Zahisham et al., 2020) [10].

Several studies have explored the application of image retrieval in e-commerce, especially for food-related websites where users search for products using images. Deep learning models like CNNs and ViTs are widely used to improve retrieval accuracy. Moreover, attention mechanisms have been incorporated to enhance feature relevance and retrieval performance, highlighting the effectiveness of focusing on specific parts of an image during retrieval tasks (Chen & Ngo, 2016) [11].

2.2 Vector Databases

Vector databases are designed for storing and querying feature vectors, making them ideal for image retrieval. Popular options include Qdrant, Milvus, Faiss, Pinecone, and

Weaviate. Milvus supports large-scale retrieval but can be complex to deploy. Pinecone is easy to scale but offers less infrastructure control. Faiss is fast with GPU support but requires technical expertise. Qdrant is chosen in this study for its balance of performance, low latency, scalability, and user-friendly documentation, making it well-suited for food image retrieval tasks.

2.3 Automatic Image Labeling

Automatic image labeling enables efficient dataset annotation, essential for training retrieval systems. Methods include weakly, semi-, and unsupervised learning. Weak supervision uses limited labeled data but depends on initial quality. Semi-supervised approaches combine labeled and unlabeled data, balancing cost and accuracy, while unsupervised methods cluster data without labels, though less accurate. Recent work highlights temporary labeling and deep learning models like CNNs and ViTs for improved retrieval. Hybrid strategies combining labeling methods have shown enhanced accuracy and scalability.

2.4 30VNFood

While global datasets like Food-101 and UEC-Food256 are widely used in food recognition, they lack detailed representation of Vietnamese cuisine. These datasets focus on broader categories or specific regions, such as Western or Japanese dishes, and often miss the subtle distinctions in Vietnamese food. In contrast, 30VNFood is designed specifically for Vietnamese cuisine, addressing challenges like visually similar dishes (e.g., phở vs. bún bò) and overlapping ingredients. This makes 30VNFood more suitable for enhancing recognition accuracy in both academic research and local e-commerce platforms, as illustrated in Fig. 1.

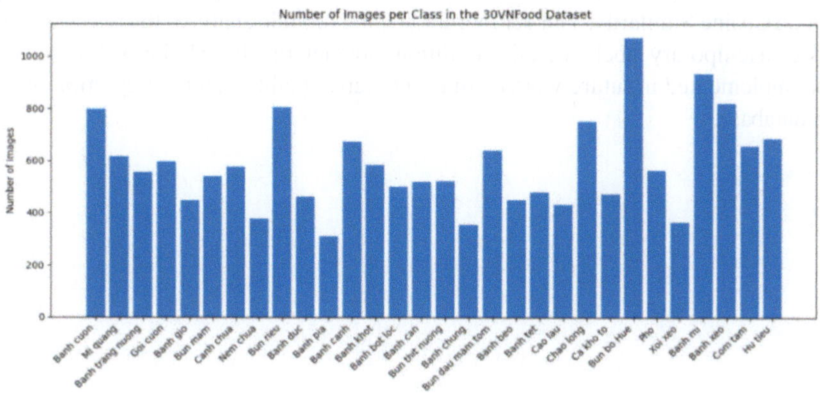

Fig. 1. Number of Images per Class in the 30VNFood Dataset

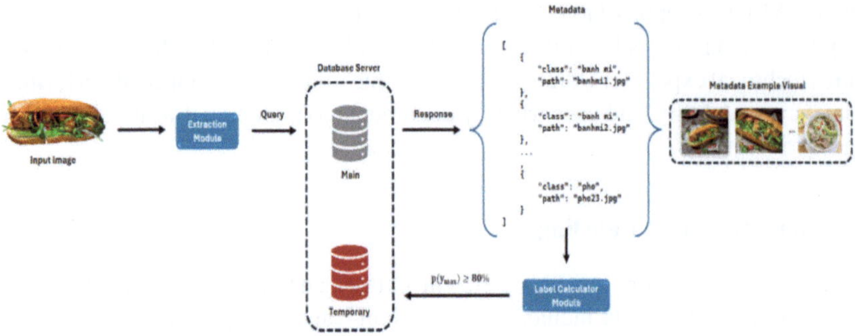

Fig. 2. Pipeline Workflow for Automatic Food Image Labeling System

3 Methodology

3.1 Pipeline Workflow

The pipeline begins with the Extraction Module, where the input image is processed using a pre-trained Vision Transformer (ViT) to extract high-dimensional feature embeddings. These features are queried against the Main Database, retrieving the Top-*k* most similar images. The retrieved metadata, including image paths and class labels, is passed to the Label Calculator Module, which analyzes the proportion of the most frequent label y_{max} among the Top-*k*. If $p(y_{max}) \geq 80\%$, y_{max} is assigned as a temporary label and stored in the Temporary Database.

The proposed pipeline for automatic food image labeling consists of four main stages (see Fig. 2): Feature Extraction, Temporal Labeling, Label Validation, and Database Integration. The workflow begins with the Extraction Module, which processes user-uploaded food images to extract high-dimensional feature embeddings. These features are then queried against the Main Database, retrieving the Top-k most similar images based on Cosine Similarity. The Temporal Labeling module analyzes the retrieved labels to assign a temporary label if certain conditions are met. Finally, a Label Validation stage (to be implemented in future work) will ensure label quality before integration into the main database.

3.2 Feature Extraction

The feature extraction module uses Vision Transformer (ViT) to produce robust embeddings for food images, enabling effective retrieval and labeling. ViT captures both local and global dependencies through self-attention and position embeddings. Unlike CNNs, ViT treats images as sequences of 16 × 16 patches, flattened and processed through attention layers. Using ImageNet pre-trained weights, ViT generalizes well without fine-tuning. The process involves resizing, normalizing, patching, and embedding images, with outputs stored in a vector database. ViT serves as the module's core, with experiments confirming its superiority over CNNs in accuracy and mAP.

3.3 Temporal Labeling

Following feature extraction, the system retrieves the Top-k most similar images from the Main Database using Cosine Similarity. The similarity between the query image and a database image is calculated as follows:

$$Sim\left(f_q, f_{d_i}\right) = \frac{f_q.f_{d_i}}{\|f_q\| \, \|f_{d_i}\|} \tag{1}$$

Where f_q is the feature vector of the query image, and f_{d_i} is the feature vector of a database image.

The labels associated with the retrieved Formally, the rule for assigning a temporary label is defined as: images are analyzed to determine the most frequent label, y_{max} The proportion of this label within the Top-k results is calculated as:

$$p(y_{max}) = \frac{Count(y_{max})}{k} \tag{2}$$

The system assigns y_{max} as a temporary label for the query image if $p(y_{max}) \geq \theta$, where $\theta = 80\%$. Formally, the rule for assigning a temporary label is defined as:

$$\hat{y} = \begin{cases} y_{max}, & if\ p(y_{max}) \geq 80\% \\ Unlabeled, & otherwise \end{cases} \tag{3}$$

This approach ensures that temporary labels are only assigned when there is a strong consensus among the retrieved results, significantly reducing the risk of incorrect labeling.

3.4 Label Validation

The Label Validation process ensures the quality of temporary labels, as demonstrated in Fig. 3, before integrating them into the main dataset. After new images are stored in the Temporal Database for about one month, they are merged with the original dataset and used to retrain a lightweight CNN model for faster updates. Following retraining, the system generates Grad-CAM visualizations to highlight the model's attention areas, allowing human reviewers to verify the accuracy of the assigned labels. Once validated, the data is extracted using ViT (Vision Transformer) pre-trained on ImageNet and merged into the Main Dataset to maintain consistent and high-quality embeddings. This approach combines automation and manual verification to ensure reliability and efficiency.

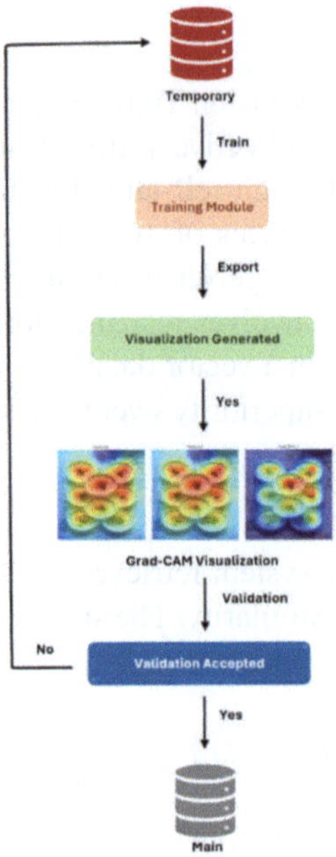

Fig. 3. Workflow with Grad-CAM Integration

4 Results and Discussion

4.1 Experimental Setting

The experiments were conducted on a high-performance setup with an AMD Ryzen 9 7900X CPU, 64 GB RAM, and an NVIDIA RTX 4090 GPU, using Python 3.9 and PyTorch. ViT and CNN models pre-trained on ImageNet were evaluated on the 30VNFood dataset. Feature embeddings were stored in Qdrant for efficient retrieval. The system was trained with a batch size of 32 and learning rate of 0.001. Performance was assessed using Precision@k and mAP, ensuring consistent and reproducible results.

4.2 Experimental Results

To evaluate system performance, various CNN and Transformer-based models were tested for feature extraction, all pre-trained on ImageNet. Metrics used included Precision@30 and mAP. As shown in Table 1, Transformer models—especially ViT-Large and ViT-Base (Patch 16)—outperformed all CNN-based models. ViT-Large achieved the highest mAP of 0.8334, followed by ViT-Base with 0.8172. In contrast, the best CNN model, EfficientNet-B7, only reached 0.7012 mAP, while others like ResNet50 and VGG16 performed even lower, highlighting their limitations with complex food datasets.

Table 1. Model Comparison on Embedding Quality

Model	Top-k	Precision	mAP
efficientnet_b0	30	0.4241	0.7038
efficientnet_b1	30	0.4281	0.7056
efficientnet_b2	30	0.4259	0.7069
efficientnet_b3	30	0.4222	0.7051
efficientnet_b4	30	0.3927	0.6995
efficientnet_b5	30	0.3831	0.6997
efficientnet_b6	30	0.3635	0.6937
efficientnet_b7	30	0.3789	0.7012
resnet101	30	0.3607	0.6847
resnet152	30	0.3696	0.6889
resnet18	30	0.2616	0.6612
resnet26	30	0.3202	0.6719
resnet34	30	0.2838	0.6637
resnet50	30	0.3250	0.6716
vgg16	30	0.3036	0.6729
vgg19	30	0.3044	0.6733
vit_base_patch16_224	30	0.6604	0.8172
vit_huge_patch14_224	30	0.6063	0.8003
vit_large_patch16_224	30	0.6910	0.8334

4.3 Discussion

The results confirm that Vision Transformers (ViT) outperform CNNs in food image labeling, with ViT-Large achieving an mAP of 0.8334 compared to 0.7012 from EfficientNet-B7. ViT's self-attention and position embeddings capture both local and global features, enhancing retrieval performance. Although ViT generalizes well without fine-tuning, its high computational cost poses challenges for real-time use. Future improvements should focus on optimizing model efficiency and database quality.

5 Conclusion

This study introduced an automated food image labeling system using ViT for feature extraction. ViT-Large and ViT-Base outperformed CNNs in Precision@k and mAP, thanks to their ability to capture both local and global features. While the system is accurate and generalizes well without fine-tuning, future work will address computational cost, database quality, and explore hybrid methods for better scalability.

References

1. Xie, Q., Hovy, E., Luong, M., Le, Q.: Self-training with noisy student improves ImageNet classification. In: Proceedings of the IEEE/CVF Conference on Computer Vision and Pattern Recognition (CVPR), Seattle, pp. 10684–10695. IEEE (2020)
2. Bilen, H., Vedaldi, A.: Weakly supervised deep detection networks. In: Proceedings of the IEEE Conference on Computer Vision and Pattern Recognition (CVPR), Las Vegas, pp. 2846–2854. IEEE (2016)
3. Qiu, J., Lo, F.P.W., Sun, Y., Lo, B.: Mining discriminative food regions for accurate food recognition. In: 30th British Machine Vision Conference (BMVC), Cardiff, pp. 1–13. BMVA Press (2019)
4. Matsuda, Y., Yanai, K.: Multiple-food recognition considering co-occurrence employing manifold ranking. In: Proceedings of the 21st International Conference on Pattern Recognition (ICPR2012), Tsukuba, pp. 2017–2020. IEEE (2012)
5. Bossard, L., Guillaumin, M., Van Gool, L.: Food-101 – mining discriminative components with random forests. In: Fleet, D., Pajdla, T., Schiele, B., Tuytelaars, T. (eds.) ECCV 2014, LNCS, vol. 8694, pp. 446–461. Springer, Cham (2014). https://doi.org/10.1007/978-3-319-10599-4_29
6. Thai, V.-P., Dang, T., Pham, Q., Nguyen, B.: Vietnamese food recognition using convolutional neural networks. In: Proceedings of the 9th International Conference on Information Science and Applications (ICISA), Seoul, pp. 124–129. Springer (2017)
7. Dalal, N., Triggs, B.: Histograms of oriented gradients for human detection. In: Proceedings of the IEEE Conference on Computer Vision and Pattern Recognition (CVPR), vol. 1, San Diego, pp. 886–893. IEEE (2005)
8. Dosovitskiy, A., Beyer, L., Kolesnikov, A., et al.: An image is worth 16x16 words: transformers for image recognition at scale. In: 9th International Conference on Learning Representations (ICLR), pp. 1–13. OpenReview.net (2021)
9. Tan, M., Le, Q.: EfficientNet: rethinking model scaling for convolutional neural networks. In: Chaudhuri, K., Salakhutdinov, R. (eds.) Proceedings of the 36th International Conference on Machine Learning, pp. 6105–6114. PMLR (2019)
10. Zahisham, Z., Lee, C.P., Lim, K.M.: Food recognition with ResNet-50. In: 2020 IEEE 2nd International Conference on Artificial Intelligence in Engineering and Technology (ICAIET), Kota Kinabalu, pp. 1–5. IEEE (2020)
11. Chen, J., Ngo, C.-W.: Deep-based ingredient recognition for cooking recipe retrieval. In: Proceedings of the 24th ACM International Conference on Multimedia, Amsterdam, pp. 32–41. ACM (2016)

Time Series-Based Electrical Device Classification on Edge with TinyML

Tolga Reis[1,2](✉) and Ahmet Teoman Naskali[1]

[1] Galatasaray University, 34349 Istanbul, Turkey
tolga.reis@siemens.com, anaskali@gsu.edu.tr
[2] Siemens A.S R&D Center, 34870 Istanbul, Turkey

Abstract. Time series classification (TSC) is crucial for edge computing and IoT, enabling intelligent decision-making in resource-constrained environments. Traditional electrical device classification methods, such as Non-Intrusive Load Monitoring (NILM), rely on handcrafted features and centralized processing, leading to latency, high computational costs, and privacy concerns. This study introduces a TinyML-based deep learning framework for real-time electrical device classification directly on edge devices. By leveraging lightweight neural networks, our approach eliminates the need for manual feature extraction while ensuring energy efficiency, scalability, and low-power execution on microcontrollers. TinyML's ability to run deep learning models locally minimizes cloud dependency, reduces inference latency, and enhances data privacy. To validate our approach, we constructed a custom time series dataset capturing detailed power consumption patterns. Experimental results indicate that TinyML-powered classification achieves robust, efficient, and real-time performance in edge-based TSC. These findings position TinyML as a scalable and energy-efficient alternative for edge-based TSC, effectively addressing NILM's limitations while unlocking new possibilities in real-time intelligent computing on embedded systems.

Keywords: Edge AI · TinyML · Signal Processing · Time Series Classification

1 Introduction

The rapid proliferation of the Internet of Things (IoT) and edge computing has led to an increasing demand for efficient and intelligent time series classification (TSC) methods. Electrical device classification is a critical application in this domain, enabling energy monitoring, fault detection, and smart automation. As global energy demand continues to rise, effective energy management has become a key priority across residential, commercial, and industrial sectors. Accurate classification and monitoring of electrical devices play a crucial role in optimizing energy consumption, reducing operational costs, and enhancing overall system efficiency.

Traditional approaches, such as Non-Intrusive Load Monitoring (NILM), rely on signal decomposition and handcrafted feature extraction [1]. However, these methods often struggle with overlapping device signatures, high computational complexity, and

A. Ito (Ed.): ICMSSP 2025, LNNS 1637, pp. 67–80, 2025.
https://doi.org/10.1007/978-3-032-05994-9_7

limited scalability in real-world environments [2]. Additionally, the increasing number of connected devices and the shift toward decentralized energy systems demand more robust, scalable, and low-latency classification techniques [3]. This necessitates the development of advanced, edge-compatible solutions that can operate efficiently in resource-constrained environments while maintaining high classification accuracy [4].

With the advancement of machine learning and deep learning techniques, automated classification methods have gained significant traction [5]. Conventional machine learning models require extensive feature engineering and cloud-based processing, leading to increased latency and dependence on centralized systems. In contrast, TinyML, a subfield of machine learning optimized for ultra-low-power microcontrollers, offers a promising alternative for performing deep learning inference directly on edge devices. This approach minimizes energy consumption, reduces cloud dependency, and enhances real-time decision-making capabilities [6].

In this study, we present a deep learning-based TinyML framework for electrical device classification on edge devices using time series data. Unlike traditional NILM-based and feature-engineered approaches [7], our method leverages lightweight neural network architectures to improve classification accuracy and computational efficiency. Additionally, we construct a custom time series dataset, ensuring realistic device usage scenarios for model training and evaluation. Our proposed solution integrates deep learning with TinyML to enable low-latency, energy-efficient classification, making it ideal for IoT-enabled smart homes, industrial monitoring, and energy-efficient applications.

The remainder of this paper is structured as follows: Sect. 2 presents the data collection process, detailing the use of Shelly devices, the types and frequency of data collected, data transmission methods, and the subsequent labeling and dataset preparation. Section 3 describes our proposed methodology, including model selection and optimization techniques. Section 4 presents experimental results, including analyses of accuracy, latency, and energy consumption. Finally, Sect. 5 concludes the paper and outlines potential future research directions.

2 Data Collection Framework

The data collection framework in this study is built upon the previous work on device recognition from electrical signals using TinyML [8]. This study introduces key improvements in data structuring, real-time inference, and model efficiency, enhancing scalability and reducing computational overhead. The optimized framework ensures seamless integration of edge-based monitoring with cloud-based processing, making it more suitable for real-world applications.

2.1 Setup

The framework consists of a Shelly Pro 3EM energy monitoring device for real-time electrical data acquisition and an ESP32 microcontroller for TinyML-based inference [9]. AWS cloud services, including IoT Core, Lambda, and S3, are used for data transmission, processing, and storage. Compared to previous implementations, this study refines data preprocessing and transmission protocols to improve latency and robustness, ensuring a more responsive and efficient pipeline. The data collection framework is detailed in Fig. 1.

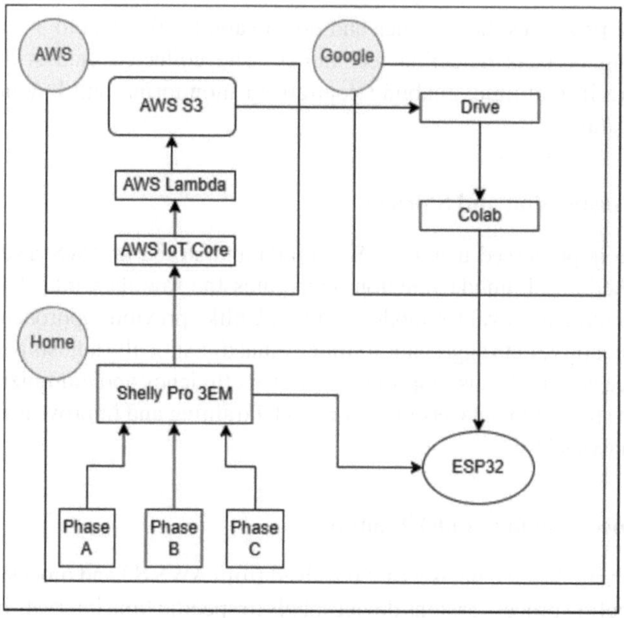

Fig. 1. System Design

2.2 Data Acquisition

The Shelly Pro 3EM device records electrical parameters, including timestamp, current (A), voltage (V), active power (W), apparent power (VA), and power factor, at a 1 Hz frequency. The device measures current using dedicated current transformers (CTs) for each phase (CTA, CTB, CTC) and optionally for the neutral line (CTN). These transformers detect the magnetic field generated by the current flow and convert it into a low-level AC voltage, which is then digitized using an Analog-Digital Converter (ADC). Voltage measurements are performed by directly connecting to the phase and neutral lines, utilizing voltage divider circuits and ADCs.

Shelly Pro 3EM samples current and voltage at a high frequency, typically in the kilohertz range, before processing the data to derive meaningful electrical parameters. The active power (W) is calculated as:

$$P = V \times I \times cos(\theta) \quad (1)$$

Where V is the instantaneous voltage, I is the instantaneous current, θ is the phase angle between them.

The apparent power (VA) is given by:

$$S = V \times I \quad (2)$$

The power factor is determined using the formula:

$$PF = P / S = cos(\theta) \quad (3)$$

The device processes these values and down-samples the data to a 1 Hz frequency using averaging or peak detection techniques. The collected data is transmitted to AWS IoT Core in real-time, enabling centralized monitoring and low-latency access to structured datasets.

2.3 Data Transmission and Storage

Real-time data is processed using AWS Lambda and stored in AWS S3 for long-term storage. A scheduled Lambda function aggregates the raw data into daily structured datasets, optimizing retrieval for model training. Unlike previous approaches, this study introduces an adaptive data aggregation strategy that dynamically adjusts data structuring based on device activity levels, improving retrieval efficiency and minimizing redundant storage. These structured datasets facilitate model training and improve accessibility for subsequent analysis [10].

2.4 Data Processing for Model Training

For model training, labeled datasets are retrieved from AWS S3 and processed in Google Colab. The labeling process assigns device labels to specific time intervals in a structured dataframe format. The time-series data is then segmented into 30-second windows, creating feature sets that include current, voltage, active power, apparent power, and power factor.

2.5 Model Training and Deployment

Once the dataset is prepared, deep learning models are trained within Google Colab with deep learning architectures designed for device classification tasks. After training, the model is optimized and converted into a TinyML-compatible format using TensorFlow Lite. To accommodate the resource constraints of ESP32, the model undergoes compression and quantization, ensuring efficient execution on the microcontroller.

Upon completion of the conversion process, the optimized model is deployed to ESP32, where it performs real-time inference. The microcontroller receives live energy consumption data from Shelly Pro 3EM, processes it through the deployed deep learning model, and classifies electrical devices accordingly. This edge-based classification reduces reliance on cloud computing, minimizes latency, and enhances system efficiency by enabling local decision-making.

2.6 Summary of Data Flow

The data collection process begins with Shelly Pro 3EM capturing real-time electrical measurements, which are immediately transmitted to AWS IoT Core. The incoming data is processed by AWS Lambda and stored in AWS S3, where it is organized into daily datasets. When required for training, the data is retrieved from AWS S3 and transferred to Google Drive for processing. Within Google Colab, the dataset undergoes labeling and time-series segmentation, preparing it for deep learning model training. The trained model is subsequently converted into a TinyML-compatible format and deployed to ESP32, which performs real-time inference on new incoming energy data. By integrating cloud-based data storage with edge-based inference, the proposed system ensures low-latency classification while maintaining scalability and computational efficiency.

3 Methodology

3.1 Dataset Description

The dataset consists of 42353 time-series sequences, each representing a 30-scond window of electrical measurements from various household appliances. Each sample is labeled with a corresponding device category, making the dataset suitable for supervised learning tasks, particularly in time-series classification and appliance recognition based on power consumption patterns.

There is a three-dimensional structure with a shape of (42353, 30, 5). The first dimension represents the total number of samples, the second dimension corresponds to the 30 time steps per sample, and the third dimension contains the 5 recorded electrical features for each time step. The dataset labels are stored separately, with a shape of (42353,), where each sample has a unique class label corresponding to the appliance in use.

The dataset includes nine distinct appliance categories, each represented by a unique integer label. The distribution of samples across these categories is as follows: 8943 samples for Airfryer (label 0), 940 samples for Blender (label 1), 2540 samples for Blow Dryer (label 2), 1864 samples for Coffee Machine (label 3), 13015 samples for Dishwasher (label 4), 9690 samples for Dryer (label 5), 2555 samples for Oven (label 6), 490 samples for Shaker (label 7), and 2316 samples for Tea Machine (label 8). Each sample in the dataset is associated with one of these appliance labels.

3.2 Data Preprocessing

The raw energy consumption data collected from Shelly Pro 3EM undergoes several preprocessing steps before being used for deep learning model training. These steps include data segmentation, cleaning and normalization, and one-hot encoding of labels, ensuring the dataset is well-structured and suitable for classification.

Dataset Splitting. The dataset is divided into three subsets: training, validation, and test sets, to ensure effective model training, hyperparameter tuning, and performance evaluation. The final training set consists of 27105 samples, which are used to train the model. A separate validation set of 6777 samples is used for fine-tuning the model and selecting the best hyperparameters [11]. The test set remains unchanged, containing 8471 samples, and is used exclusively for evaluating the model's final performance on unseen data. This dataset split provides a balanced approach, allowing the model to learn from the training data, adjust based on validation performance, and be assessed on a completely independent test set for unbiased evaluation.

Cleaning and Normalization. To enhance the model's ability to distinguish between different appliances, the existing electrical parameters in the dataset are segmented into 30-second windows and labeled accordingly [12]. Each feature set is inherently structured within the dataset and undergoes preprocessing to ensure consistency. To maintain numerical stability, Min-Max Scaling is applied, and missing values, positive infinite values, and negative infinite values are replaced with zero to prevent numerical inconsistencies during training.

One-Hot Encoding. The class labels are converted into a one-hot encoded format, where each appliance category is represented as a binary vector of length 9. Each vector contains a single "1" at the index corresponding to the active device and "0" in all other positions [13].

3.3 Model Architecture and Training

The training process for all models follows a structured approach to optimize performance and ensure robust generalization. Each model is trained using supervised learning, where the input consists of 30 time steps with 5 recorded electrical features per step, and the output is a one-hot encoded class label representing the active appliance.

During training, the Adam optimizer is used with its default learning rate of 0.001, as it provides efficient convergence by adapting learning rates based on gradient estimates [14]. The loss function employed is categorical cross-entropy, which is well-suited for multi-class classification problems. To enhance model generalization and prevent overfitting, early stopping is implemented with a patience of 5 epochs, meaning training will stop if the validation loss does not improve for 5 consecutive epochs. Additionally, ReduceLROnPlateau dynamically reduces the learning rate by a factor of 0.5 when the validation loss stagnates for 3 consecutive epochs, allowing for more refined weight updates in later epochs. ModelCheckpoint is also used to save the best-performing model during training, ensuring that the most optimal version is retained.

The models are trained for a maximum of 50 epochs, but the actual training duration may be shorter if early stopping is triggered. A batch size of 32 is used, balancing computational efficiency and convergence stability.

Fully Convolutional Network (FCN). The FCN model consists of three convolutional layers, each with batch normalization and ReLU activation for stable training and feature extraction. The first layer (128 filters, kernel size 8) captures long-range dependencies, the second (256 filters, kernel size 5) refines features, and the third (128 filters, kernel size 3) focuses on fine-grained details.

Unlike traditional networks, FCN replaces fully connected layers with Global Average Pooling (GAP), reducing parameters while preserving temporal information, enhancing generalization, and minimizing overfitting [15]. The final Dense layer (9 units) applies softmax activation for classification. The model has 270,985 parameters (1.03 MB), with 269,961 trainable and 1,024 non-trainable parameters due to batch normalization.

Time Convolutional Neural Network (TimeCNN). The TimeCNN model consists of three convolutional layers with batch normalization and ReLU activation for stable feature extraction. The first layer (64 filters, kernel size 3) captures short-range dependencies, the second (128 filters, kernel size 5) refines mid-level features, and the third (256 filters, kernel size 7) captures long-range dependencies [16].

GAP reduces each feature map to a single value, minimizing parameters while preserving learned representations. The final Dense layer (9 units) applies softmax activation for classification. The model has 275,849 parameters (1.05 MB), with 274,953 trainable and 896 non-trainable due to batch normalization.

Residual Network for Time-Series Classification (ResnetID). The ResNet-ID model captures complex temporal dependencies using residual connections, preventing the vanishing gradient problem. Unlike traditional convolutional models, it employs skip connections to retain important features [17].

It starts with a convolutional layer (64 filters, kernel size 3), followed by batch normalization and ReLU activation. It then includes three residual blocks, each with two convolutional layers (kernel size 3) and increasing filter sizes (32, 64, 128). Skip connections and batch normalization enhance feature preservation and stability. A 1×1 convolution ensures dimensional alignment when needed.

GAP reduces feature map dimensionality while retaining key representations. A Dense layer (128 units) processes features before the final output layer (9 units, softmax activation) classifies appliance categories. The model has 135,081 parameters (527.66 KB), with 134,057 trainable and 1,024 non-trainable due to batch normalization.

Temporal Convolution Network (TCN). The TCN model is designed for time-series classification using causal and dilated convolutions to capture long-range dependencies efficiently [18]. Unlike standard CNNs, it expands the receptive field exponentially without significantly increasing parameters, making it effective for sequential data.

It consists of three residual blocks with 1D convolutions at dilation rates of 1, 2, and 4. Causal padding ensures each time step is influenced only by past values, preserving temporal structure. Each block includes two convolutional layers (64 filters, kernel size 3), batch normalization, and dropout (rate = 0.3). Skip connections with 1x1 convolutions align dimensions and prevent vanishing gradients.

After residual blocks, GAP reduces feature maps, and a Dense layer (9 units, softmax) classifies appliance categories. The model has 73,609 parameters (287.54 KB), with 72,841 trainable and 768 non-trainable due to batch normalization.

Squeeze-and-Excitation Time-Series Network (SE-TSNet). The SE-TSNet model improves time-series classification by integrating Squeeze-and-Excitation (SE) blocks into a 1D CNN, dynamically recalibrating channel-wise features to enhance selectivity and robustness [19].

It consists of three convolutional layers: 32 filters (kernel size 3) for short-term dependencies, 64 filters (kernel size 5) for mid-range dependencies, and 128 filters (kernel size 7) for long-term dependencies. Each layer is followed by an SE block that applies GAP and a squeeze-and-excitation mechanism using a sigmoid activation to refine feature importance.

GAP further reduces feature dimensions while preserving key information before the final Dense layer (9 units, softmax) classifies appliance categories. The model has 75,077 trainable parameters (293.27 KB). While SE blocks slightly increase complexity, they improve feature selectivity and generalization, making SE-TSNet highly effective for time-series classification.

Simple Recurrent Neural Network (SimpleRNN). The SimpleRNN model is designed for time-series classification, leveraging recurrent connections to capture sequential dependencies [20]. Unlike convolutional models, it processes data step by step, preserving temporal relationships.

It consists of two SimpleRNN layers: the first (64 units) returns sequences for further processing, while the second (32 units) outputs the final hidden state. Both layers use tanh activation and L2 regularization (0.001) to prevent overfitting.

Batch normalization follows each recurrent layer, and dropout (rate = 0.3) enhances generalization. The final Dense layer (9 units, softmax) performs multi-class classification. The model has 8,265 parameters (32.29 KB), with 8,073 trainable and 192 non-trainable due to batch normalization.

3.4 Model Optimization

The model optimization process was applied uniformly to minimize memory usage and improve inference efficiency, making models suitable for edge devices and TinyML applications. TensorFlow Lite post-training quantization was used to reduce model size and enhance computational performance [21].

The optimization included integer-only quantization, converting weights and activations to INT8 for low-power hardware compatibility. Dynamic range quantization stored weights in reduced precision while allowing dynamic activation adjustments, calibrated using a representative dataset from real test data [22].

The process involved converting trained models to TensorFlow Lite format and applying INT8 quantization, ensuring a fully optimized TFLite model. This approach improves inference speed while maintaining classification accuracy, enabling efficient deployment on resource-constrained edge devices.

3.5 Edge Inference

The Edge Inference process deployed the optimized quantized model on an ESP32 microcontroller for real-time execution in a resource-constrained environment. The fully quantized TensorFlow Lite model was converted into a C++ header file, storing model weights and structure in a format compatible with embedded systems [23].

Once deployed, the ESP32 loaded the model, preprocessed test data to match its quantization format, and performed real-time predictions. The system successfully classified input sequences without relying on cloud computing, proving its feasibility for on-device appliance recognition.

This approach ensures low-latency inference, reduces power consumption, and enhances data privacy by keeping computations local. The successful ESP32 deployment confirms the model's efficiency for real-world TinyML applications.

4 Experimental Results

4.1 Performance of Standard Models

The performance of the standard models was evaluated using the test dataset before applying quantization. The evaluation included key classification metrics such as accuracy, precision, recall, and F1-score (classification metrics), all of which were computed using weighted averages to ensure that the impact of each class was proportional to its frequency in the dataset. Additionally, the total model size was recorded to assess feasibility for deployment in memory-constrained environments. The results are summarized in Table 1.

Among all models, ResNet-ID achieved the highest accuracy (99.95%), along with the best classification metrics, demonstrating superior feature extraction capabilities through its residual connections while maintaining a relatively small model size of 1.68 MB [24]. SE-TSNet also performed exceptionally well (99.88%), benefiting from the squeeze-and-excitation mechanism, which enhances feature representations while keeping the model size minimal at 0.94 MB, the smallest among convolution-based architectures.

FCN and SimpleRNN delivered competitive accuracy levels of 98.55% and 98.60%, respectively. However, their architectural differences are significant. FCN, with a model size of 3.16 MB, utilizes fully convolutional layers, making it effective for time-series feature extraction. In contrast, SimpleRNN achieved similar accuracy but with the smallest model size (0.14 MB), making it the most lightweight model. However, its sequential nature may limit its ability to capture long-range dependencies as effectively as convolution-based approaches.

TimeCNN exhibited slightly lower accuracy (96.52%), possibly due to its fixed kernel sizes, which might have reduced its adaptability to varying time dependencies. However, its classification metrics remain competitive, showing that it still generalizes well across different appliance categories.

TCN performed well (97.08%) while having the second smallest model size (0.97 MB), highlighting the efficiency of temporal convolutions with dilated connections. This balance between performance and efficiency suggests that TCN is a strong alternative for time-series classification, especially in scenarios where model size is a critical constraint.

Table 1. Test Performance of Standard Models.

Model	Accuracy (%)	Precision	Recall	F1-Score	Model Size (MB)
FCN	98.55	98.55	98.55	98.53	3.16
TimeCNN	96.52	96.58	96.52	96.42	3.21
ResnetID	99.95	99.95	99.95	99.95	1.68
TCN	97.08	97.22	97.08	97.10	0.97
SE-TSNet	99.88	99.88	99.88	99.88	0.94
SimpleRNN	98.60	98.60	98.60	98.56	0.14

4.2 Performance of Quantized Models

After applying post-training quantization, the models were evaluated on the test dataset to assess their classification performance and efficiency in a memory-constrained environment. The evaluation included key classification metrics. Additionally, the model size after quantization was recorded to measure the reduction in memory usage. The results are summarized in Table 2.

Among the quantized models, SE-TSNet achieved the highest accuracy (99.32%), maintaining strong performance even after quantization while significantly reducing its model size to 98.38 KB, the smallest among all architectures. FCN also performed well (98.39%), showing minimal accuracy degradation compared to its floating-point counterpart while reducing its size from 3.16 MB to 282.84 KB.

TimeCNN and SimpleRNN delivered competitive accuracy levels of 96.29% and 97.65%, respectively, with their quantized models showing only a slight drop in performance. SimpleRNN retained its strong classification capability (97.65%) while keeping its model size minimal at 192.98 KB, making it an efficient option for deployment in resource-limited environments.

ResNet-ID exhibited a more noticeable performance drop, decreasing from 99.95% to 94.75% accuracy, suggesting that its residual connections might be more sensitive to quantization effects. However, its model size was significantly reduced to 177.59 KB, making it highly efficient in terms of memory consumption.

TCN showed the largest accuracy drop (from 97.08% to 87.71%), indicating that its dilated convolutional structure was affected by quantization. However, it remained the second smallest model at 107.87 KB, highlighting its potential as a lightweight alternative despite its performance reduction.

The results in Table 2 show that SE-TSNet and FCN retained the highest classification performance after quantization, while SimpleRNN and TCN demonstrated efficient memory usage with relatively small model sizes. ResNet-ID experienced a more significant performance drop, despite its reduced size, suggesting that additional optimizations may be needed to enhance its robustness in quantized form. The next section analyzes the performance of these models on edge devices, focusing on inference speed and memory utilization.

Table 2. Test Performance of Quantized Models.

Model	Accuracy (%)	Precision	Recall	F1-Score	Model Size (KB)
FCN	98.39	98.41	98.39	98.38	282.84
TimeCNN	96.29	96.40	96.29	96.19	285.51
ResnetID	94.75	94.90	94.75	94.70	177.59
TCN	87.71	89.81	87.71	88.05	107.87
SE-TSNet	99.32	99.38	99.32	99.33	98.38
SimpleRNN	97.65	97.63	97.65	97.63	192.98

4.3 Performance on the Edge Device

The quantized models were deployed on an ESP32 to evaluate their efficiency in a resource-constrained environment. The ESP32 used in this study features two CPU cores running at 240 MHz, 355 KB of total heap memory, and 4 MB of flash memory, making it suitable for low-power AI applications. The inference process was conducted using the TFLiteMicro library, which allows efficient execution of TensorFlow Lite models on microcontrollers with limited computational resources [25]. Two key factors were analyzed: memory usage, which refers to the amount of RAM occupied by the model during inference, and inference time, which represents the time required to process a single input and generate a prediction. The results are summarized in Table 3.

Table 3. Test Performance on the Edge Device.

Model	Memory Usage (KB)	Inference Time (Sec)
FCN	20.65	0.53
TimeCNN	20.17	0.53
ResnetID	24.54	0.23
TCN	19.7	0.09
SE-TSNet	13.82	0.14
SimpleRNN	33.18	0.01

Memory Usage. One of the critical constraints in edge computing is memory availability, as microcontrollers have significantly less RAM compared to traditional computing devices. The results show that SE-TSNet had the lowest memory usage at 13.82 KB, making it highly suitable for deployment on ESP32. TCN also demonstrated strong memory efficiency, consuming only 19.7 KB, making it one of the most lightweight convolutional architectures.

On the other hand, SimpleRNN had the highest memory usage at 33.18 KB, which can be attributed to its recurrent nature, requiring additional storage for maintaining hidden states. ResNet-ID consumed slightly more memory (24.54 KB), likely due to its residual connections, while FCN and TimeCNN had similar memory footprints (around 20 KB), reflecting their comparable computational structures.

Inference Speed and Real-Time Processing. Inference speed is a crucial factor for real-time applications, where rapid decision-making is required. SimpleRNN demonstrated the fastest inference time at 0.01 s per sample, making it highly efficient for low-latency applications. TCN followed with an inference time of 0.09 s, indicating that its dilated convolutions are well-optimized for sequential data processing on low-power hardware.

Among the convolutional architectures, SE-TSNet had a relatively fast inference time of 0.14 s, showing that the squeeze-and-excitation mechanism did not introduce significant computational overhead. ResNet-ID had an inference time of 0.23 s, while FCN and TimeCNN required 0.53 s per sample, making them the slowest models in this comparison.

5 Conclusion and Future Work

5.1 Conclusion

This study evaluated multiple deep learning models for time-series classification of electrical appliance usage, focusing on both standard floating-point performance and quantized model efficiency for edge deployment on an ESP32 microcontroller. The results demonstrated that quantization significantly reduces model size while maintaining high classification accuracy, making deep learning feasible for low-power embedded systems.

The results showed that SE-TSNet and ResNet-ID achieved the best classification accuracy, with SE-TSNet demonstrating high efficiency in memory usage after quantization. SimpleRNN was the fastest model in terms of inference speed but required the most memory, limiting its applicability in highly constrained devices. FCN and TimeCNN showed strong classification performance but had the slowest inference times on the ESP32, making them less suitable for real-time applications. TCN provided a balance between accuracy, memory efficiency, and inference time, making it a competitive choice for real-world deployment on low-power edge devices.

The deployment of TinyML models for real-time appliance classification shows that deep learning can be applied to low-cost microcontrollers without relying on cloud-based solutions. However, optimizing models for both accuracy and efficiency remains a challenge, as different architectures offer trade-offs between classification performance, memory footprint, and real-time processing capabilities.

5.2 Future Work

While the quantized models demonstrated successful deployment on an ESP32 microcontroller, further improvements can be explored to enhance both accuracy and efficiency. One key direction for future work is the application of federated learning, which enables decentralized model training across multiple edge devices without transmitting raw data to a central server. This approach can improve privacy, reduce communication overhead, and allow continuous learning from real-world device usage patterns.

Another optimization strategy involves model pruning and knowledge distillation to further reduce model size while maintaining classification performance. Pruning techniques can remove redundant parameters, making models even more suitable for ultra-low-power microcontrollers. Knowledge distillation, where a smaller model learns from a larger pre-trained model, could also be investigated to balance inference speed and accuracy.

Additionally, hybrid quantization techniques, combining post-training quantization and quantization-aware training, could help mitigate accuracy degradation observed in certain models after quantization. Exploring alternative quantization schemes such as mixed-precision quantization may provide further improvements in efficiency without sacrificing performance.

Expanding this work to other embedded platforms with more computational resources, such as dedicated AI accelerators and low-power neural processing units (NPUs), could further improve inference speed and enable more complex models to run efficiently on edge devices. Investigating the integration of TinyML-based anomaly detection alongside appliance classification may also offer additional real-world applications, particularly in energy efficiency monitoring and predictive maintenance.

By addressing these challenges, future research can push the boundaries of on-device deep learning, making TinyML models even more adaptive, scalable, and capable of running autonomously in real-world environments.

References

1. Hart, G.W.: Nonintrusive appliance load monitoring. Proc. IEEE **80**(12), 1870–1891 (1992)
2. Zoha, A., Gluhak, A., Imran, M.A., Rajasegarar, S.: Non-intrusive load monitoring approaches for disaggregated energy sensing: a survey. Sensors **12**(12), 16838–16866 (2012)
3. Shi, W., Cao, J., Zhang, Q., Li, Y., Xu, L.: Edge computing: vision and challenges. IEEE Internet Things J. **3**(5), 637–646 (2016)
4. Li, E., Zhou, Z., Chen, X.: Edge intelligence: on-demand deep learning model co-inference with device-edge synergy. In: Proceedings of the 2018 Workshop on Mobile Edge Communications (MECOMM), Budapest, pp. 31–36. ACM (2018)
5. Ouyang, H., Xie, S., Chen, M., Cao, X., Wang, Y.-N., Yuan, M.: A method based on LSTM network for detection of abnormal daily electricity. In: Proceedings of the 2021 6th International Conference on Multimedia Systems and Signal Processing (ICMSSP), New York, pp. 48–52. ACM (2021)
6. Alajlan, N.N., Ibrahim, D.M.: TinyML: Enabling of inference deep learning models on ultra-low-power IoT edge devices for AI applications. Micromachines **13**(6), 851 (2022)
7. Ruzzelli, A.G., Nicolas, C., Schoofs, A., O'Hare, G.M.P.: Real-time recognition and profiling of appliances through a single electricity sensor. In: International Workshop on Agent-Oriented Software Engineering, 2010 Proceedings, pp. 1–10. IEEE (2010)
8. Reis, T., Naskali, A.T.: Device recognition from electrical signals with TinyML. Trans. Comput. Sci. Appl. **1**(1) (2024)
9. Zim, M.Z.H.: TinyML: analysis of Xtensa LX6 microprocessor for neural network applications by ESP32 SoC. arXiv preprint arXiv:2106.10652 (2021)
10. Borra, P., Mullapudi, M., Nerella, H., Prakashchand, L.: Analyzing AWS edge computing solutions to enhance IoT deployments. Int. J. Eng. Adv. Technol. (IJEAT) **13**(6), 8–14 (2024)
11. Kohavi, R.: A study of cross-validation and bootstrap for accuracy estimation and model selection. In: IJCAI (1995)
12. Keogh, E., Chu, S., Hart, D., Pazzani, M.: Segmenting time series: a survey and novel approach. In: Data Mining in Time Series Databases, pp. 1–22. World Scientific (2004)
13. Poslavskaya, E., Korolev, A.: Encoding categorical data: Is there yet anything 'hotter' than one-hot encoding? arXiv preprint arXiv:2312.16930 (2023)
14. Kingma, D.P., Ba, J.: Adam: a method for stochastic optimization. arXiv preprint arXiv:1412.6980 (2014)
15. Wang, Z., Yan, W., Oates, T.: Time series classification from scratch with deep neural networks: a strong baseline. arXiv preprint arXiv:1611.06455 (2017)
16. Sawada, A., Miyagawa, T., Ebihara, A.F., Yachida, S., Hosoi, T.: Convolutional neural networks for time-dependent classification of variable-length time series. arXiv preprint arXiv:2207.03718 (2022)
17. Ismail Fawaz, H., Forestier, G., Weber, J., Idoumghar, L., Muller, P.-A.: Deep learning for time series classification: a review. Data Min. Knowl. Discov. **33**(4), 917–963 (2019). https://doi.org/10.1007/s10618-019-00619-1
18. Pelletier, C., Webb, G.I., Petitjean, F.: Temporal convolutional neural network for the classification of satellite image time series. arXiv preprint arXiv:1811.10166 (2018)

19. Hu, J., Shen, L., Sun, G.: Squeeze-and-excitation networks. In: Proceedings of the IEEE Conference on Computer Vision and Pattern Recognition (CVPR), Salt Lake City, pp. 7132–7141. IEEE (2018)
20. Smirnov, D., Nguifo, E.M.: Time series classification with recurrent neural networks. In: Proceedings of the 3rd ECML/PKDD Workshop on Advanced Analytics and Learning on Temporal Data, pp. 1–9 (2018)
21. Jacob, B., et al.: Quantization and training of neural networks for efficient integer-arithmetic-only inference. arXiv preprint arXiv:1712.05877 (2017)
22. Hubara, I., Nahshan, Y., Hanani, Y., Banner, R., Soudry, D.: Improving post training neural quantization: layer-wise calibration and integer programming. arXiv preprint arXiv:2006.10518 (2020)
23. Zhou, C., et al.: MicroNets: neural network architectures for deploying TinyML applications on commodity microcontrollers. arXiv preprint arXiv:2010.11267 (2020)
24. He, K., Zhang, X., Ren, S., Sun, J.: Deep residual learning for image recognition. In: Proceedings of the IEEE Conference on Computer Vision and Pattern Recognition (CVPR), Las Vegas, pp. 770–778. IEEE (2016)
25. Schneider, D., et al.: Deploying quantized neural networks on microcontrollers: the TFLite micro approach. In: Proceedings of the 2023 ACM/IEEE International Conference on Internet of Things Design and Implementation (IoTDI), San Antonio, pp. 45–56. IEEE (2023)

Domain-Specific Image Captioning: Vietnamese Cuisine on the 30VNFoods Dataset

Huynh Vu Nhu Nguyen(✉), Khang Nguyen Hoang, Khac Huy Hynh, and Hoang Ngoc Tran

FPT University, Cantho, Vietnam

{huynhnvnce170550,khangnhce171197,huyhkce171229}@fpt.edu.vn

Abstract. This study evaluates models on the 30vnfood dataset to optimize Vietnamese food captioning. We propose a pipeline combining CLIP and MBart, achieving high BLEU (0.9313), CIDEr (7.4063), and METEOR (0.9311) scores. Integrating vision and language features proved essential for generating accurate captions. Our results provide actionable insights for food e-commerce and VNLP by addressing the challenges of a low-resource language in image captioning tasks. The evaluation of model performance on the Vietnamese food dataset, 30vnfood, is a critical task in determining optimal models for specific objectives. In this study, we propose a novel pipeline combining CLIP-based image embeddings with a trainable MBart model for Vietnamese caption generation. The pipeline utilizes a frozen CLIP model for feature extraction, a linear projection layer for dimensionality alignment, and MBart for natural language generation. This research systematically evaluates the effectiveness of the proposed architecture in generating captions for Vietnamese food images. Our results provide insights into the model's performance, highlighting its potential for broader applications in food-related industries and automated systems.

Keywords: Vietnamese food captioning · CLIP · MBart · image captioning · VNLP · low-resource language · e-commerce applications

1 Introduction

Recent progress in machine learning has greatly advanced image captioning, particularly in domain-specific contexts. However, generating accurate and descriptive captions in Vietnamese for food images remains a relatively unexplored challenge. The 30vnfood dataset—comprising a rich and diverse set of Vietnamese culinary images—provides an ideal resource for developing and evaluating models in this niche.

This research is conducted as part of an undergraduate thesis project aiming to build an e-commerce platform centered on Vietnamese cuisine. The platform seeks to improve user engagement by incorporating automated image captioning to deliver meaningful descriptions for food items.

To address this task, we propose a novel pipeline that integrates CLIP, a powerful vision-language model, with MBart, a multilingual sequence-to-sequence transformer.

A. Ito (Ed.): ICMSSP 2025, LNNS 1637, pp. 81–87, 2025.
https://doi.org/10.1007/978-3-032-05994-9_8

A trainable linear projection layer is employed to align CLIP's image embeddings with MBart's input space for effective caption generation in Vietnamese.

The primary objective is to identify an optimal and computationally efficient model architecture for domain-specific captioning. Our work not only evaluates the performance of this integration but also contributes to Vietnamese Natural Language Processing (VNLP) and low-resource captioning research. Ultimately, the findings have practical implications for automated food recognition, recommendation systems, and digital cultural preservation.

2 Related Work

MBart, a multilingual transformer model, has shown significant potential in sequence-to-sequence tasks due to its ability to handle diverse language structures [1]. Designed to facilitate machine translation across multiple languages, MBart has been widely applied in tasks such as translation and summarization [2, 3]. Its architecture supports the integration of different language pairs, making it a versatile tool in multilingual natural language processing (NLP). Models such as mBART-50 have further advanced these capabilities, but their application in domain-specific tasks like image captioning, particularly for low-resource languages such as Vietnamese, remains limited [4].

The challenge of integrating visual and textual modalities in low-resource languages has prompted the exploration of transformer-based architectures in both computer vision and NLP. In the field of computer vision, transformers such as Vision Transformer (ViT) and Detection Transformer (DETR) have demonstrated superior performance in tasks like image classification, object detection, and segmentation [5]. ViT, which processes image patches as input tokens, has enabled models to capture long-range dependencies, while DETR's end-to-end approach eliminates the need for traditional region proposals [6]. These advancements have laid the groundwork for vision-language models capable of generating high-quality captions by effectively combining visual and textual embeddings.

In image captioning, OSCAR and VinVL have established themselves as benchmarks by leveraging object tags to align visual-text representations. OSCAR improves visual-text alignment using object semantics, while VinVL incorporates richer vision datasets to further enhance performance [7, 8]. These models highlight the advantages of integrating robust visual embeddings with rich textual representations, a key aspect of generative tasks like image captioning. However, these advances have primarily focused on English-centric datasets, leaving Vietnamese applications relatively underexplored.

For Vietnamese NLP tasks, progress has been comparatively slower due to the lack of large annotated datasets and the language's unique grammatical complexity. Models like PhoBERT have played a pivotal role in addressing these challenges by providing robust language representations for Vietnamese [9]. PhoBERT has proven effective in various NLP tasks, but its integration into multimodal applications such as image captioning has yet to be fully explored. Although some efforts, such as Nguyen et al.'s introduction of a Vietnamese-specific image-text dataset, have addressed data scarcity, most of these works have focused on retrieval tasks rather than generative captioning.

Recent works have highlighted the potential of using vision-language models like CLIP to bridge these gaps. CLIP, trained on large-scale vision and language data, has shown remarkable performance in generating robust visual embeddings. Our approach builds on these advancements by integrating CLIP with MBart, leveraging CLIP's visual representation capabilities and MBart's multilingual text generation strengths. This integration enables our pipeline to generate syntactically and semantically accurate captions for Vietnamese food images, addressing the challenges posed by low-resource datasets. Furthermore, fine-tuning on the 30vnfood dataset ensures that the generated captions are contextually relevant and domain specific [11].

Our proposed methodology aligns with existing research trends in low-resource image captioning and vision-language modeling. Works such as EfficientNet and Swin Transformer emphasize the importance of efficient model scaling and hierarchical representations in vision tasks [12, 13]. Additionally, recent advances in cross-lingual pretraining and self-supervised learning offer promising directions for improving the generalizability of models in multilingual and multimodal contexts [14, 15]. By building on these studies, our research aims to further the development of automated food image captioning systems while contributing to the broader field of Vietnamese NLP and computer vision.

3 Methodology

3.1 Dataset

Fig. 1. Examples of dishes from the 30vnfood dataset with they caption created with format.

The 30vnfood dataset is a curated collection of high-resolution images showcasing traditional and modern Vietnamese dishes. It supports image captioning in low-resource languages by capturing diverse regional cuisines. Each image is paired with a standardized Vietnamese caption in the format *"món ăn trong hình là {class_name}"*, ensuring clarity and consistency. The high-quality visuals enable accurate caption generation and make the dataset ideal for applications like food recognition, menu digitization, and recommendation systems. Examples of the dataset are shown in Fig. 1.

3.2 Pipeline Design

Our proposed methodology utilizes a multi-stage pipeline to generate Vietnamese captions for food images in the 30vnfood dataset. The pipeline consists of three key components: a frozen CLIP image encoder model, a trainable linear projection layer, and a fine-tuned MBart model for Vietnamese language generation, as illustrated in Fig. 2.

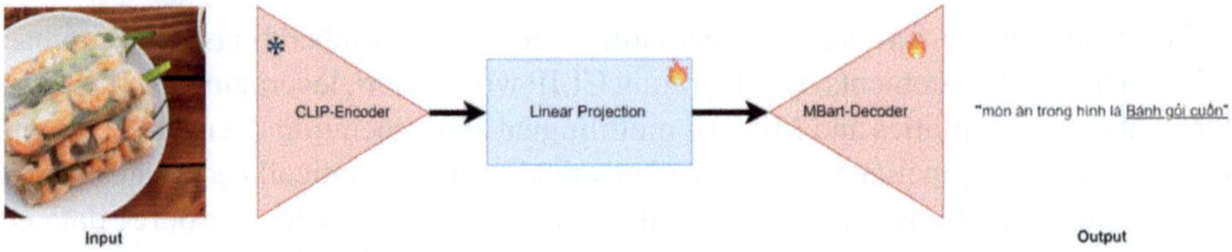

Fig. 2. Pipeline Design for Vietnamese Caption Generation Using CLIP and Mbart

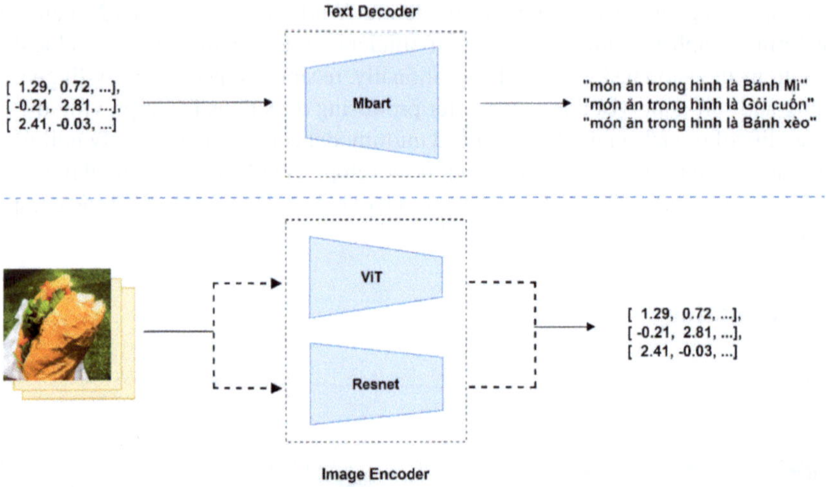

Fig. 3. Illustration of the pipeline design showcasing the input image processed through the image encoder and text decoder components

CLIP, a contrastive vision-language model pre-trained on large-scale image-text pairs, plays a crucial role in extracting rich, high-dimensional image embeddings. By leveraging its frozen parameters, CLIP ensures robust and consistent feature extraction, which captures both the semantic content and visual characteristics of the food images. This guarantees that the embeddings retain meaningful information across diverse input samples, as shown in Fig. 3.

The extracted image embeddings are passed through a trainable linear projection layer, which aligns the dimensionality of the CLIP embeddings with the input space of MBart, facilitating smooth integration. This alignment is essential for bridging the gap between the visual representation provided by CLIP and the linguistic generation capabilities of MBart.

Finally, the MBart model, a multilingual sequence-to-sequence transformer, generates Vietnamese captions through fine-tuning on the 30vnfood dataset. This process minimizes cross-entropy loss between predicted and ground truth captions, ensuring accurate and contextually relevant outputs. To further enhance model performance and prevent overfitting, we applied various optimization techniques, including learning rate scheduling, early stopping, and dropout.

3.3 Evaluation Metrics

To evaluate the quality of the generated captions, we employed several standard metrics widely used in image captioning tasks: BLEU, CIDEr, METEOR, and ROUGE. Each metric captures distinct aspects of caption quality, providing a comprehensive assessment of syntactic accuracy, semantic relevance, and contextual coverage. The BLEU (Bilingual Evaluation Understudy) metric measures the n-gram overlap between generated captions and reference captions, emphasizing fluency and syntactic accuracy. It is calculated as follows:

$$BLEU = BP.exp\left(\sum_{n=1}^{N} w_n logp_n\right) \tag{1}$$

where BP is the brevity penalty to account for length differences, p_n denotes the precision of n-gram matches, and w_n is the weight for each n-gram level. A higher BLEU score indicates closer alignment with the reference captions.

CIDEr (Consensus-based Image Description Evaluation) focuses on contextual relevance by assigning higher importance to terms that frequently appear in reference captions but are rare in the dataset, using Term Frequency-Inverse Document Frequency (TF-IDF) weighting. The CIDEr score is computed as:

$$CIDEr = \frac{1}{N}\sum_{i=1}^{N} \frac{\sum_{j=1}^{M} w_{ij}.w'_{ij}}{\sqrt{\sum_{j=1}^{M}\left(w_{ij}\right)^2.\sqrt{\sum_{j=1}^{M}\left(w'_{ij}\right)^2}}} \tag{2}$$

Where w_{ij} and w'_{ij} represent the TF-IDF weighted frequencies of term j in the predicted and reference captions, respectively. This metric highlights the model's ability to generate captions that capture the key visual details of an image for Evaluation of Translation with Explicit ORdering) evaluates the semantic alignment of captions, considering synonyms, stemming, and word order. Its formulation is:

$$METEOR = F_{mean}.(1 - Penalty) \tag{3}$$

Where F_{mean} is the harmonic mean of precision and recall, and Penalty accounts for disordered phrase matches. A high METEOR score reflects accurate and contextually appropriate captions.

Finally, ROUGE (Recall-Oriented Understudy for Gisting Evaluation) measures the overlap of n-grams or sequences of words between the generated and reference captions. For ROUGE-N, the score is defined as:

$$ROUGHE - N = \frac{\sum_{S \in SetRef} \sum_{gram \in S} Count_{match}\left(garm_N\right)}{\sum_{S \in SetRef} \sum_{gram \in S} Count\left(garm_N\right)} \tag{4}$$

where N is the size of the n-gram. ROUGE-1 and ROUGE-2 capture unigram and bigram overlaps, while ROUGE-L focuses on the longest matching sequence. High ROUGE scores indicate effective recall of essential textual elements.

Together, these metrics provide a holistic evaluation framework, ensuring the generated captions are syntactically accurate, semantically relevant, and contextually aligned with the reference captions.

4 Results and Discussion

4.1 Experimental Setting

Experiments were run on a high-performance system with an AMD Ryzen 9 7900X CPU, 64 GB RAM, and an NVIDIA RTX 4090 GPU. The pipeline, combining CLIP for image embedding and MBart for text generation, was fine-tuned on the 30vnfood dataset using Python 3.12, PyTorch, and HuggingFace Transformers. Optimization techniques like learning rate scheduling and early stopping were applied to ensure efficiency and prevent overfitting, resulting in high performance across all evaluation metrics.

4.2 Experimental Results

Table 1 summarizes the performance of three different CLIP variants: clip-vit-large-patch14, clip-vit-base-patch32, and clip-vit-base-patch16 used in our proposed pipeline. The evaluation was conducted using standard image captioning metrics: BLEU, CIDEr, METEOR, and ROUGE. For consistency, the MBart-large-50-many-to-many-mmt model was kept unchanged across all configurations.

Table 1. Comparison of CLIP Variants

Metric	clip-vit-large-patch14	clip-vit-base-patch32	clip-vit-base-patch16
BLEU	0.8693	0.8744	0.9313
CIDEr	6.6558	6.2940	7.4063
METEOR	0.8713	0.8757	0.9311
ROUGE1	0.8922	0.8910	0.9327
ROUGE2	0.8687	0.8677	0.9225
ROUGE-L	0.8922	0.8907	0.9325

Among the three configurations, clip-vit-base-patch16 consistently outperformed the others across all metrics. This model achieved a BLEU score of 0.9313, indicating its superior syntactic alignment with the reference captions. Additionally, its CIDEr score of 7.4063 reflects its strong contextual relevance, emphasizing critical terms in the captions. The model also achieved a high METEOR score of 0.9311, demonstrating its ability to maintain semantic consistency.

The ROUGE metrics further confirm the effectiveness of clip-vit-base-patch16 in accurately capturing essential textual elements. Specifically, the model achieved scores of 0.9327 (ROUGE-1), 0.9225 (ROUGE-2), and 0.9325 (ROUGE-L), showcasing its robustness in generating comprehensive and contextually aligned captions.

4.3 Discussion

The results highlight that clip-vit-base-patch16 significantly outperforms other CLIP variants, confirming its strength in capturing detailed visual and semantic features crucial for Vietnamese image captioning. While other models performed adequately, they lacked alignment with MBart's language generation. The strong performance of this pipeline demonstrates its suitability for low-resource languages like Vietnamese, especially in contexts with limited annotated data and cultural complexity. The integration of CLIP and MBart proves effective, enabling real-world applications such as automated product descriptions, digital menus, and food recommendation. Future work will aim to expand the dataset and improve generalizability through self-supervised and cross-lingual learning.

5 Conclusion

This study proposed a novel pipeline combining CLIP and MBart for Vietnamese food image captioning. The clip-vit-base-patch16 variant yielded the best performance on the 30vnfood dataset. This integration demonstrates strong potential for automated applications in Vietnamese e-commerce and cultural preservation. Future work will focus on expanding the dataset and optimizing the model for broader domain-specific use cases.

References

1. HuggingFace: MBart Model Documentation. https://huggingface.co/transformers/v3.5.1/model_doc/mbart.html. Accessed 13 Jan 2025
2. Liu, Y., et al.: RoBERTa: a robustly optimized BERT pretraining approach. arXiv preprint, arXiv:2001.08210 (2020)
3. Facebook AI: MBart-large-50-many-to-many-mmt Model. https://huggingface.co/facebook/mbart-large-50-many-to-many-mmt. Accessed 13 Jan 2025
4. Nguyen, T.T., Nguyen, Q.T., Le, D.C.: A dataset for Vietnamese image-text retrieval tasks. arXiv preprint, arXiv:2210.05610 (2022)
5. Dosovitskiy, A., et al.: An image is worth 16x16 words: transformers for image recognition at scale. arXiv preprint, arXiv:2010.11929 (2020)
6. Carion, N., Massa, F., Synnaeve, G., Usunier, N., Kirillov, A., Zagoruyko, S.: End-to-end object detection with transformers. arXiv preprint, arXiv:2005.12872 (2020)
7. Li, X., Yin, X., Li, C., Hu, X., Zhang, P., Zhang, L., et al.: OSCAR: object-semantics aligned pre-training for vision-language tasks. arXiv preprint, arXiv:2004.06165 (2020)
8. Zhang, Z., et al.: VinVL: making visual representations matter in vision-language models. arXiv preprint, arXiv:2101.00529 (2021)
9. Nguyen, T., Vu, T., Pham, V.: PhoBERT: pre-trained language model for Vietnamese. arXiv preprint, arXiv:2003.00744 (2020)
10. Radford, A., et al.: Learning transferable visual models from natural language supervision. arXiv preprint, arXiv:2103.00020 (2021)
11. 30vnfood Dataset Repository. https://github.com/ds4v/30VNFoods. Accessed 13 Jan 2025
12. Tan, M., Le, Q.V.: EfficientNet: rethinking model scaling for convolutional neural networks. In: Proceedings of ICML, pp. 6105–6114 (2019)
13. Liu, Z., Lin, Y., et al.: Swin transformer: hierarchical vision transformer using shifted windows. In: Proceedings of ICCV, pp. 10012–10022 (2021)
14. Conneau, A., Lample, G.: Cross-lingual language model pretraining. In: NeurIPS, pp. 7059–7071 (2019)
15. He, K., Fan, H., et al.: Momentum contrast for unsupervised visual representation learning. arXiv preprint, arXiv:1911.05722 (2020)

Multimodal Signal Recognition and Classification

Using Artificial Intelligence for Identification of Disaster Tweets with an Improved Deep Learning Approach

Hamad Zogan(✉), Khan Mohamed Jarina Begum, Vidhya Sivalingam, and Khalid Hasan Mahmoud Alsinjlawi

Department of Computer Science, Jazan University, Jazan, Kingdom of Saudi Arabia
hzogan@jazanu.edu.sa

Abstract. Artificial intelligence (AI) offers astonishing capabilities, with current achievement in using human experience at different phases of classical building. Only when AI actively collaborates with humans to co-create solutions as part of its intelligence will it reach its full potential. This study can use data convergence to combine AI simulation and human intelligence to increase decision-making as well as create a capability like to "teaming intelligence." The study of information extraction from tweets employs several techniques. It can comprehend human words and sentences due to Natural Language Processing (NLP), a subfield of AI. NLP blends statistical, deep learning (DL) as well as machine learning (ML) models with rule-based human language modelling. The goal of this study is to use NLP and pipelines for disaster tweet classification. Since tweets are by their very nature largely unstructured, text pre-processing—which entails eliminating unnecessary and extraneous words from the tweets—is a crucial step in the process. Text pre-processing, feature extraction, and modelling are all processes in the NLP pipeline. This use a TF-IDF vectorizer for pre-processing, which encompasses tokenization, stop word removal, tokenization, and feature extraction. To assess the data based on tweets, this work makes use of a hybrid classifier and compares with several ML and DL classification techniques. The proposed approach incorporates a hybrid classifier, comparing its performance against several ML and DL techniques. This study provides insights into optimizing NLP pipelines for real-world applications by evaluating the classifiers' effectiveness on disaster-related tweet data.

Keywords: Artificial intelligence · Deep learning technique · Natural language processing · Classification · Machine learning technique

1 Introduction

The occurrence of cyclones, earthquakes, floods, and wildfires has increased dramatically during the last decade on a global scale. These catastrophic disasters have the potential to claim many lives and destroy a significant amount of property, thereby jeopardizing has physical and environmental safety. Deploying a multitude of video surveillance camera systems enables the monitoring of large-scale disasters. These smart technologies can

A. Ito (Ed.): ICMSSP 2025, LNNS 1637, pp. 91–102, 2025.
https://doi.org/10.1007/978-3-032-05994-9_9

automatically evaluate and categorize catastrophe footage in real-time thanks to their well-trained AI models. Equally the accuracy of the AI outcome along with processing speed are critical for real AI model deployments to develop effective disaster response [1]. Impressive capabilities have emerged from AI, which has recently been able to successfully use human experience at different points in the model generation process. Only when AI actively collaborates with humans to co-create solutions will it reach its full potential. It is difficult for decision-makers and authorities to quickly analyze complicated information and data in order to make evacuation choices during a flood catastrophe. For this reason, it is critical to provide more precise and quick evacuation models in order to make sensible and safe decisions during emergencies [2].

Smartphones, sensors, satellites, closed-circuit televisions, drones, and other cutting-edge technology generate massive amounts of data during catastrophe situations. As mentioned earlier, the large amount, variety, and diversity of data make it challenging for emergency responders to make sense of the available information in a time-sensitive situation. On top of that, the sheer amount of data makes it impossible for first responders to do manual analyses, which is crucial for efficient and successful incident management. When it comes to processing massive amounts of varied data and extracting meaningful insights, AI is indispensable. To enhance many operations, processes, and pattern discovery in large datasets, AI refers to the research and development of software tools that may mimic human intelligence to extract high-level abstractions from large datasets [3]. AI is a division of computer science that attentions on computers to do tasks traditionally associated with human cognition, such as learning, remembering, making decisions, solving problems, etc. Since then, there has been a lot of progress in the subject, and it has found uses in many other areas [4].

An increasing number of fields are beginning to use DL, which has its roots in image and video categorization, NLP, networking, robotics, cybersecurity as well as mobile computing between many others. DL may also have a favorable effect on how humanitarian organizations and emergency response teams conduct damage assessments during natural disasters. This emergency squad can react quickly and effectively with the help of DL damage assessments, which help them identify the right resources to handle the aftermath. The capacity to make predictions using computers and mathematical models is a key component of DL, which has strong ties to computational statistics [5]. An emergency is defined as any circumstance that has the potential to suddenly endanger human life. At any moment, with no advance notice, an emergency may strike. Being well prepared is the best way to handle these unforeseen accidents and disasters. By using geographical targeting in SMS notifications, users may avoid panicking people in unaffected regions while still informing those in areas impacted by an occurrence. In times of crisis, such as natural disasters, SMS warnings are a lifesaver and help get people to safety faster. Ongoing emergency management initiatives include using an SMS alert system. Any member may use this system [6].

Social media real-time data, including text, photos, and videos, can be a valuable source of information for disaster response and detection. But a critical difficulty arises from the huge and intricate nature of social media data: how to efficiently and precisely extract important information about disasters from these enormous datasets. Several obstacles currently confront the field of disaster management. The problem of information overload comes first. The exponential rate of social media information sharing makes it difficult to filter out relevant content, as most of it is noise and unrelated to disasters. Second, unlike traditional sources of knowledge about disasters, social media provides a very diverse and unstructured source of information. Furthermore, disaster information is highly timely, and any delay could have dire consequences. As a result, it is critical to develop a technique that can quickly and precisely identify disaster-related material from social media in order to increase disaster response efficacy and efficiency [7]. These disaster characteristics influence decision-making on the ground, and social media content's applicability in assisting these decision-making processes for responses across various sectors is always evolving. Therefore, as operations transition from response to recovery, annotation activities on social media content for disaster data analytics systems may encounter inconsistent word and context relevance for response sectors, leading to a loss of significance [8].

This paper's remaining sections center on this structure. Section 2 provides a summary of relevant studies on the application of AI techniques in disaster management. Section 3 describes the proposed methodology. Section 4 discusses the outcomes along with a comparative analysis. The references are found after Sect. 5's discussion of the conclusion.

2 Recent Works for Research

This section examines some of the most recent techniques based on AI techniques. Table 1 details several novel techniques. Present methods include a suggested technique, including AI, NLP, CNN and DL techniques. These technologies provide an advantage, including improved performance. Nevertheless, other downsides, such as only a limited number of messages were evaluated and to improve the data availability and quality.

Amrit Sen et al. [9] created a software-based prototype that can identify dangers from ambient sound and automatically notify victims' registered contacts via email, SMS, or WhatsApp without the use of any additional hardware. They use Exploratory Data Analytics (EDA) techniques to visualize and analyze audio signals from the Kaggle dataset. They achieve this by supplying various DL models with EDA outcomes. When identifying audio events, CNNs with long short-term memory (LSTM) provide an accuracy of 96.6%. In addition to providing the user with three choices for predicting the class of the live audio, the ability to choose several models becomes useful as the dataset and recorded audio get more complicated and larger.

Zair Bouzidi [10] put forth a strategy that opens the way for the integration of social media, DL, AI technology, and the detection or avoidance of fake news into the crisis management paradigm. In order to efficiently extract content, this model uses a hybrid of Deep CNN and LSTM as its main tool. CNN collects meaningful features from various sources to obtain qualitative as well as reliable data, while LSTM maintains long-term dependencies in the extracted features with recurrent connections.

Ismet et al. [11] sought to evaluate the health professionals' estimated aid requirements for a week following the earthquakes in Kahramanmaraş, Turkey, as reported by those who willingly participated in the WhatsApp communication network. Volunteers

Table 1. Summary of previous works their methods, advantages and limitations

Paper and Author	Method	Advantages	Limitations
Amrit Sen et al. [9]	NLP, DL	Needs no additional wearables or hardware (such separate watches)	Not included are details on the victim's whereabouts, the nature of their distress, the severity of the emergency, vital signs (such as heart rate and oxygen levels), or the surrounding environment (such as temperature and altitude)
Zair Bouzidi [10]	Convolutional Neural Network (CNN)	Better performance	Did not use multiple languages,
İsmet et al. [11]	Whatsapp Communication Network	Explains the use of social media during a disaster	Only a limiterd number of messages were evaluated
Ayoub et al. [12]	IoT	It aids in lowering flood hazards by enhancing decision-making, preparing for evacuation, and allocating resources effectively	Has to improve the data availability and quality
Seyoung Lee et al. [13]	AI Chatbots	Chatbots' emotional expressions influence people's intentions to take action to reduce their impact on the environment in a positive way	It assessed only intentions to mitigate climate change, not actual actions

in Turkey formed a WhatsApp group following the earthquake, and this descriptive retrospective epidemiological study looks at their message activity over the course of a week. There were a total of 5,813 messages sent within the 7-day time frame that was considered for the investigation. For various reasons, the research excluded 3472 communications, leaving only 2341 messages for use in the study.

Ayoub et al. [12] described their research to build and deploy an advanced early warning and flood forecasting system that makes use of smart technology to allow for accurate predictions and preventative actions in areas prone to flooding. Their proposed system would monitor the river's water level in real time using Internet of Things (IoT) technology. They analyze the gathered data using an association rule technique to generate accurate predictions of future flood occurrences. The intelligent flood disaster prediction system will greatly benefit users and authorities by providing early warnings and enabling educated decisions on evacuation, resource allocation, and infrastructure strengthening. With the system's early flood forecasting capabilities, high-risk areas can be better prepared and respond to floods more effectively, thus saving more lives. The results of this study emphasize the system's potential to enhance flood risk management tactics and lessen the human and material costs associated with floods in at-risk areas.

Seyoung Lee et al. [13] examined how chatbots' emotional expressions affected people's intentions to take action to reduce their impact on climate change, as well as the separate as well as sequential mediation functions of social attendance along with guilt. In addition, this research examined the direct and indirect associations between the factors in question, as well as the moderating effect of custom addressing. They discovered that social attendance influenced the association among chatbots' emotional expression as well as climate-change mitigation behavioral purpose, with the former generating more intentions than the latter. Even though guilt did not play a mediating role, the data showed that social attendance and guilt could mediate in a serial fashion. The present investigation employs ML and AI methodologies to examine the following research questions:

- RQ1: How might ML and DL identify tweets in the event of a disaster?
- RQ2: How can ML and DL categorically encode several stages in a disaster?

This study aims to create and test Twitter models that are efficient at identifying and classifying tweets, with a particular emphasis on Twitter communication during significant disasters. This work makes significant contributions to the following fields:

- LSTM is a popular DL technique for text classification. Learning is possible, and LSTM takes long-term reliability into account. As a result, this paper proposes a hybrid model that combines CNN with a gated recurrent unit (GRU).
- The primary goal of this hybrid model's proposal is to combine the benefits of a very dependable CNN architecture for managing high-dimensional data with the ability to extract local features from text. The GRU, which excels in processing sequence data, combines the benefits of CNN.
- The combination of these benefits will provide an optimal classifier model, contributing to the creation of an early warning system for disaster detection.

3 Proposed Methodology

In both serious and informal contexts, Twitter has emerged as a vital means of communication. In times of crisis, people are more likely to call for assistance using their mobile phones. Disaster relief organisations are interested in keeping tabs on Twitter conversations since they have grown in popularity. In some cases, the context of a tweet is unclear. The real issue lies in identifying the tweets about the actual disaster and disregarding the rest [14]. Figure 1 shows the real-time automated disaster intelligence methodological framework.

3.1 Dataset

The researchers used 10,876 tweets as a dataset. The figure eight corporation initially made this dataset public on its website. This continued to use the dataset for Kaggle contests until the study was completed. A binary classification job is one possible way to approach this problem. If the tweet is about a real tragedy, this wants the label y = 1, and otherwise, the label y = 0.

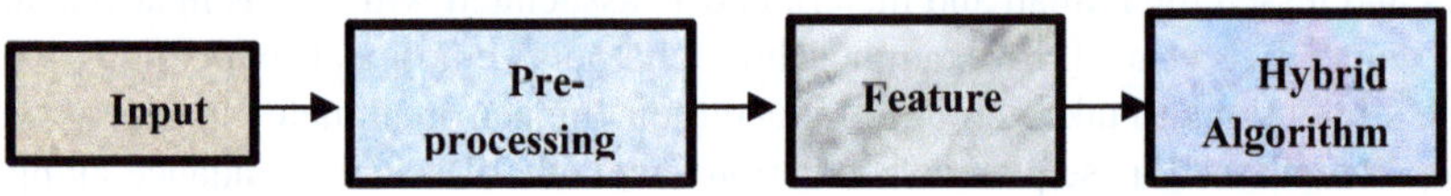

Fig. 1. Real-time automated disaster intelligence framework

3.2 Preprocessing

Twitter data is directly associated with the following: link sharing, subject highlighting using hashtags, referring to other users using the @ sign, emoticon character symbols, word abbreviations, and poorly structured sentences. It might be wise to do further research on Twitter text characteristics before moving on to the word embedding step. For many text mining methods, text preparation is a crucial first step. It usually contains tasks like stemming, tokenization, filtering, and lemmatization.

The following may be removed. It removes URL character references (e.g., https, http), HTML tags (e.g.,/p, tbody, h1, pre), and numeric values (e.g., 5 km, M194) that incorporate a blend of alphabetical characters, stopwords, hashtag mentions, and punctuation. After that, take out any character references (such as lt, amp, or nbsp) and extract any repeated characters from large words. Then, shift all of the capitalization to lowercase and use the word's context to convert it to its fundamental tenses, a process called lemmatization.

Here, it provides a dataset example that compares text before and after text preprocessing. The text processing tools have performed flawlessly, converting all words to lowercase. From the first three lines, it appears that the process of removing stopwords is going well. Similarly, this eliminates any extraneous punctuation, such as hashtags. The lemmatization approach effectively transforms all sentences in the dataset into fundamental tenses. The text pre-processing results are as follows in Table 2 [16].

3.3 Feature Extraction

The goal of feature extraction is to create numerical features from raw data while keeping the original dataset intact. This work uses the use of two functions like TF-IDF vectorizer and Glove. Applying a vectorizer to a text document generates a document term matrix, with terms in the columns and documents in the rows (sometimes referred to as a bag of words). The Tfidf Transformer is used for feature extraction approach. It stands for "term frequency times inverse document frequency," or Tfidf in Eq. (1).

$$\text{Inverse Direction Factor} = \log[(1 + D)/(1 + df(d, t))] + 1 \tag{1}$$

Table 2. Text Preprocessing result

Raw text	Cleaned text
is different cities, stay safe everyone There is a forest fire	different city stay safe
at spot pond, geese are fleeing across the street, I cannot save them all	forest fire spot pond geese flee street save
Just happened a terrible car crash Heard about #earthquake	happen terrible car crash hear earthquake

Where D represents the number of documents, and df(d, t) represents the number of times a term has occurred in the document term matrix. The aforementioned formula is used to calculate the Tf-idf scores.

The 2014 development by Stanford University of matrix factorization algorithms on word-context matrices is the basis of Glove: Global Vectors. This can use word-context matrices, a type of co-occurrence matrix, to identify the semantic relationship between words within a corpus. This used the glove word embedding tool from Kaggle, which provides a large corpus with four distinct embedding representations [15].

3.4 Proposed Hybrid Classifier

Next, it need to prepare the architecture of the classifier model. Combining the reliable GRU module for sequence data with the excellent CNN module for high-dimensional data is the main objective of this CNN-GRU hybrid model. The first layer, embedding, records the text's vector representation. After that, it uses a 1D CNN with 64 filters, a 1D kernel, the same padding, and ReLu activation to build the convolution. After that, an average pooling layer takes the 1D CNN output and uses the average to lower its dimensionality. A total of 128 neurons make up the GRU layer, which receives the average pooling output.

The GRU layer uses tanh activation and 0.2 recurrent dropouts. The RepeatVector layer transforms the 2D shape of the GRU layer into a 3D shape, while the GlobalMax-Pooling 1D layer recovers the 2D shape. Following the GlobalMaxPooling1D layer, a 0.2-size dropout layer will receive the output, followed by a dense layer with a neuron configuration twice as large as the previous layer. Each dense layer repeats three times, with the only configuration change being the number of neurons. The last layer determines whether the tweet should be classified as a disaster based on the configuration of one neuron's value and the sigmoid activation function supplied by the fully connected layer. Table 3 shows the hybrid architecture employs a total of thirteen layers in its layer order.

Table 3. Hybrid CNN-GRU proposed model summary

Layers (type)	Output Shape
Dense (fully connected)	(None, 1)
Drop Out	(None, 1024)
Dense	(None, 1024)
Drop Out	(None, 512)
Dense	(None, 512)
Drop Out	(None, 256)
Dense	(None, 256)
Global Average Pooling 1D	(None, 128)
Repeat Vector	(None, 22, 128)
Gated Recurrent Unit	(None, 128)
Average Pooling 1D	(None, 22, 64)
Convolutional 1 D	(None, 22, 64)
Embedding	(None, 22, 300)

4 Results and Discussions

For exploratory data analysis in Jupyter Notebook, this study makes use of Python. To optimize loss throughout the training phase, the Adam optimiser configuration employs a 20% test proportion, a 10-test epoch, and an auto-reduce learning rate function.

4.1 Evaluation Measures

This study uses standard measures for classification tasks, such as accuracy, precision, recall, and F1-score, to evaluate the performance of the proposed models.

Accuracy: This metric calculates the accuracy of tweet predictions in relation to the total number of predicted tweets in Eq. (2).

$$\text{Accuracy} = \frac{TP + TN}{TP + FP + FN + TN} \tag{2}$$

Precision: It is the proportion of positive predictions that are accurate in Eq. (3).

$$\text{Precision} = \frac{TP}{TP + FP} \tag{3}$$

Recall: This proportion is correctly classified; real positives make up for it in Eq. (4).

$$\text{Recall} = \frac{TP}{TP + FN} \tag{4}$$

F1-score: This harmonic mean combines precision and recall in Eq. (5).

$$\text{F} - \text{score} = 2 * \frac{precision * recall}{precision + recall} \tag{5}$$

4.2 Experimental Results

This section presents the experimental results of the proposed work in a tabular format, obtained by independently implementing the TF-IDF vectorizer and glove.

Table 4. Classification results with TF-IDF vectorizer.

Methods	Precision	Recall	F1-score	Accuracy
Proposed	85.21	84.11	84.09	84.13
Random Forest	78.06	84.91	64.35	73.11
Gradient boost	79.85	86.63	65.34	74.35
XGboost	80.23	81.58	74.91	77.45
CNN	81.41	83.86	75.47	79.99
GRU	81.54	77.17	81.31	79.05

Table 4 displays the comparison findings. The proposed approach has a recall of 84.11%, an accuracy of 84.13%, a precision of 85.21, and a f1 score of 84.09. While gradient boost, XGboost, CNN, and GRU all achieve accuracy levels below the proposed technique, random forest achieves a 15.07 percent improvement. Random Forest has a precision that is 9.15% greater than the gradient approach, whereas XGBoost, CNN, and GRU all have precisions that are lower than the proposed method. The recall for Random Forest, on the other hand, is 0.95 percentage points greater than the proposed technique, while the recall for gradient boost, XGboost, CNN, and GRU is 2.99%, 3.10 percent, and 0.29%, respectively. Finally, the proposed method's F1-score is 30% lower than Random Forest, XGboost is 12.25%, CNN is 11.42%, and GRU is 3.41%, while the GRU gradient is 28.69%. Figure 2 shows the classification results with TF-IDF vectorizer comparison graph.

Table 5 displays the comparison findings. The proposed approach has a recall of 85.01%, an accuracy of 84.56%, a precision of 86.17%, and a f1 score of 84.97%. The accuracy of random forest is 4.50% greater than the proposed technique, while that of CNN, XGboost, gradient boost, and GRU is 8.92%, 10.50%, 6.52%, and 6.17%, respectively. While gradient boost is 6.68%, XGboost is 9.46%, CNN is 6.79%, and GRU is 5.41%, random forest has a precision that is 9.64% greater than the proposed technique. The recall for Random Forest, on the other hand, is 0.16% greater than the proposed technique, compared to 3.15% for gradient boost, 7.34% for XGboost, 9.16% for CNN, and 9.37% for GRU. Random forest outperforms the proposed technique by 27.87% in terms of F1-score, compared to 15.74% for gradient boost, 13.52% for XGboost, 5.98% for CNN, and 2.62% for GRU. Figure 3 shows the classification results with glove comparison graph.

It is clear that the proposed CNN-GRU hybrid architecture might be able to classify data well enough to get the highest score, even when tested using two different pre-trained word embedding methods.

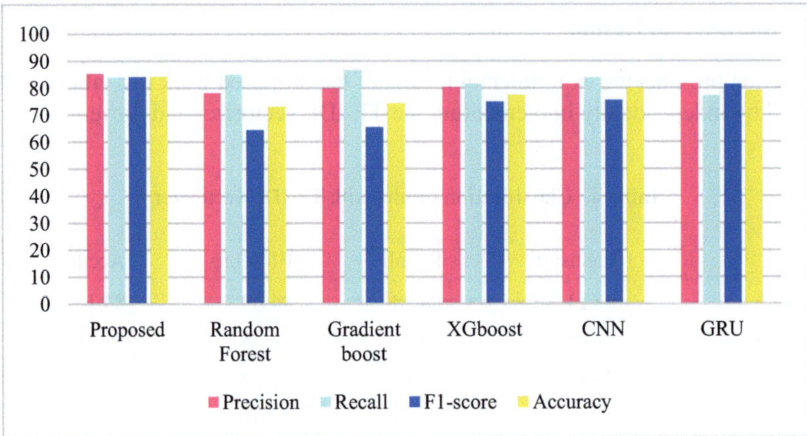

Fig. 2. Comparison graph of TF-IDF vectorizer classification

Table 5. Classification results with Glove

Methods	Precision	Recall	F1-score	Accuracy
Proposed	86.17	85.01	84.97	84.56
Random Forest	78.59	84.87	66.45	74.64
Gradient boost	80.77	82.12	73.41	77.63
XGboost	78.72	79.19	74.85	76.52
CNN	80.69	77.87	80.17	79.38
GRU	81.74	77.72	82.80	79.64

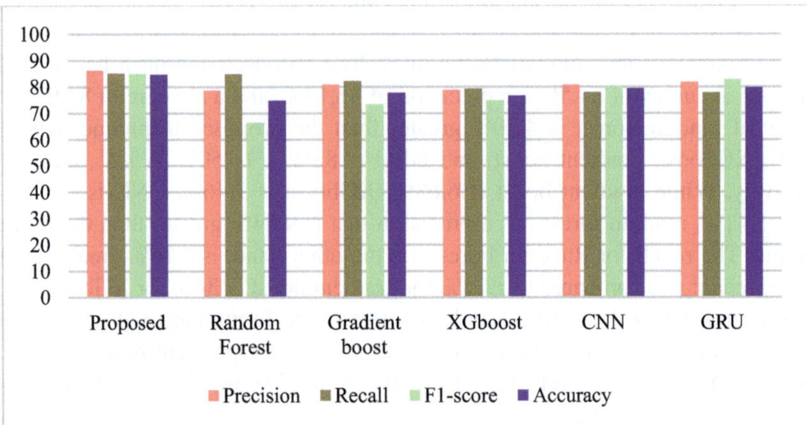

Fig. 3. Comparison graph of glove classification

5 Conclusion

To make sure the proposed model is appropriate for identifying disaster tweets, this study compares it to a variety of different classifiers. This evaluated the performance of this proposed model using both the TF-IDF vectorizer and glove approaches. All tests consistently show that the proposed model has the best classification performance, particularly in terms of accuracy and f1-score, when compared to various other classifiers. This hybrid CNN-GRU model with Glove earned an accuracy score of 86.17% and an F1-score of 84.56%. The comparative results show that Glove performs marginally better when paired with this proposed hybrid GRU rate for the dataset scenarios are working on. These findings demonstrate that the proposed hybrid model consistently outperforms its competitors when using two distinct word embedding strategies. The findings may vary, however, when using various datasets. Future testing of this CNN hybrid model should undoubtedly utilize additional text-based datasets.

References

1. Sze Yang Tan, N., Tham, M.-L., Chua, S.Y., Lee, Y.L.: Accurate and fast classification of natural disasters using CNN-LSTM and inference acceleration. J. Commun. Softw. Syst. **20**(1), 58–68 (2024)
2. Karanjit, R., Samadi, V., Hughes, A., Murray-Tuite, P., Stephens, K.: Converging Human Intelligence with AI systems to advance flood evacuation decision making. Nat. Hazards Earth Syst. Sci. Discuss., 1–29 (2024)
3. Aboualola, M., Abualsaud, K., Khattab, T., Zorba, N., Hassanein, H.S.: Edge technologies for disaster management: a survey of social media and artificial intelligence integration. IEEE Access (2023)
4. Pwavodi, J., Ibrahim, A.U., Pwavodi, P.C., Al-Turjman, F., Mohand-Said, A.: The role of artificial intelligence and IoT in prediction of earthquakes. Artif. Intell. Geosci., 100075 (2024)
5. Liang, H., Burgess, L., Liao, W., Blasch, E., Yu, W.: Deep learning assist IoT search engine for disaster damage assessment. Cyber Phys. Syst. **9**(4), 313–337 (2023)
6. Mishra, S.: Development of SMS alert module for emergency response planing in QGIS
7. Adili, P., Chen, Y.: Fast disaster event detection from social media: an active learning method. Int. J. Comput. Commun. Control **19**(2) (2024)
8. Ara, Z., et al.: Closing the knowledge gap in designing data annotation interfaces for AI-powered disaster management analytic systems. In: Proceedings of the 29th International Conference on Intelligent User Interfaces, pp. 405–418 (2024)
9. Sen, A., Rajakumaran, G., Mahdal, M., Usharani, S., Rajasekharan, V., Vincent, R., Sugavanan, K.: Live event detection for people's safety using NLP and deep learning. IEEE Access (2024)
10. Bouzidi, Z.: Deep learning-based managing disaster with smart education in environmental sustainability while avoiding fake news. Adv. Mach. Learn. Artif Intell. **5**(1), 01–18 (2024)
11. Çelebi, İ., Tuna, A., Çalişkan, C.: Evaluation of incoming messages to WhatsApp communication network created by volunteers in major disasters: the case of Turkey earthquake. Disaster Med. Public Health Preparedness **18**, e53 (2024)
12. Ayoub, N.A., Aziz, A.A., Mustafa, W.A.: Original research article Floodintel: advancing flood disaster forecasting through comprehensive intelligent system approach. J. Auton. Intell. **7**(1) (2024)

13. Lee, S., Park, Y, Park, G.: Using AI chatbots in climate change mitigation: a moderated serial mediation model. Behav. Inf. Technol., 1–17 (2024)
14. Lakshmi, S.D., Velmurugan, T.: Classification of disaster tweets using natural language processing pipeline. Acta Sci. Comput. Sci. **5**(3) (2023)
15. Krishna, D.S., Gorla, S., Prasad Reddy, P.V.G.D.: Disaster tweet classification: a majority voting approach using machine learning algorithms. Intell. Decis. Technol., 1–13 (2023)
16. Pratama, R.A.M., Pardede, H.F.: Disaster tweet classifications using hybrid convolutional layers and gated recurrent unit. Int. J. Comput. Digit. Syst. **14**(1), 1 (2023)

Generation of Listening Motion of Embodied Conversational Agents Using Speech and Text Information

Haruki Ito, Akinori Ito(✉), and Takashi Nose

Graduate School of Engineering, Tohoku University, Sendai 980-8679, Japan
aito.spcom@tohoku.ac.jp

Abstract. This research focuses on generating nonverbal behavior when an embodied conversational agent (ECA) is the listener. While previous studies on ECAs have focused on generating actions when the agent is the speaker, research on generating actions as a listener has been limited. In particular, it is known that in Japanese dialogue, unlike English dialogue, the correlation between speech and head movements is low. Therefore, a model that considers both linguistic and non-linguistic information is needed. This study constructed a model based on Learning2Listen (L2L) and considered the speaker's past movements and speaker ID in addition to the speaker's speech and text information. To improve the model's performance, VQ-VAE was improved by introducing exponential moving average (EMA) and code reset. In the objective evaluations, VQ-VAE with EMA and code reset showed better performance, and multimodal learning using speech and text was shown to be effective. In the subjective evaluation, the human-likeness of the generated movements and their appropriateness as a listener were evaluated. As a result, the proposed method was shown to be superior to randomly generated movements in terms of appropriateness as a listener. However, compared to actual human movements, the proposed method was suggested to have room for improvement.

Keywords: Conversational agent · listening motion · VQ-VAE · Spoken dialogue

1 Introduction

Conversational agents are systems that provide information and task support by engaging in natural dialogue with the user through speech and text, such as a smart speaker or a communication robot. Conversational agents integrate speech recognition, natural language processing, speech synthesis, dialogue control, and other technologies to interact with the user. In particular, virtual characters and robots with faces and bodies capable of non-verbal expressions such as facial expressions and body movements are called Embodied Conversational Agents (ECAs) [1]. For such ECAs to be used in everyday life, it is necessary not only to improve the naturalness of responses but also to provide nonverbal interactions such as facial expressions and gestures so that users can become more immersed in the dialogue.

A. Ito (Ed.): ICMSSP 2025, LNNS 1637, pp. 103–112, 2025.
https://doi.org/10.1007/978-3-032-05994-9_10

Many methods for generating nonverbal behaviors for anthropomorphic dialogue agents have been studied so far. In the early days of research, the method [2, 3] was considered to generate behaviors by using a rule-based system. In rule-based systems, movements are designed by hand and are expected to be natural and easy for humans to understand. However, since many rules must be developed by hand, it is challenging to accommodate diverse patterns of human behavior, and the system is prone to repetitive actions in similar situations. For this reason, deep learning-based approaches have become mainstream in recent years.

The GENEA Challenge is a collaborative effort to build a gesture generation system. The most highly rated model in GENEA2023 is the Diffusion-based Model [4]. This system ranked first among the submitted systems regarding humaneness and speech adequacy. However, it was ranked second from the bottom for interlocutor adequacy because it did not include any information about the interlocutor. The other models also scored very low in interlocutor adequacy compared to the other two measures, and the generation of actions when the ECA is the listener has not been thoroughly investigated.

In English dialogue, several studies have been on the generation of listener actions [5–9]. Jonell et al. [5] proposed a method to generate facial gestures that the interlocutor influences. Ng et al. [7] also proposed a model using VQ-VAE and generates listener movements autoregressively from speaker movements and speech.

In Japanese dialogue, several studies have examined the generation of actions by ECAs during utterances [10–13]. For example, Ishii et al. [13] proposed a method for generating nonverbal actions such as nodding, head posture, facial expressions, hand gestures, and upper body posture by inputting words, parts of speech, word positions, and dialogue acts. However, generation of actions when the dialogue agent is the listener was not considered.

The studies of listening motion generation for English dialogues use only the interlocutor's speech; however, it is known that the correlation between speech and head motion is low in Japanese dialogue. For example, a study investigating the correlation between head movements and F0 was 0.39–0.52 for English speakers and 0.22–0.30 for Japanese speakers. A study examining the relationship between linguistic and nonlinguistic information in spoken utterances showed that nodding, shaking the head sideways, and face-up movements were often associated with linguistic information such as aids, fillers, negations, questions, and phrase boundaries, and were less often associated with nonlinguistic information such as prosody [14].

In this study, we model the nonverbal behavior of a dialogue agent as a listener using a combination of speech and text information as input and investigate the effect of each modality on the precision of prediction. This study's model is based on Learning2Listen (L2L) [7] and extended to allow speaker ID and text input. VQ-VAE is also improved to enhance model performance. The learned generative models are evaluated using objective evaluation metrics.

Another issue in motion generation for Japanese dialogues is the dataset. To the best of our knowledge, there is no large-scale dialogue data that includes motion capture data. To solve this problem, a method has been proposed to perform 2D human posture estimation from video using OpenPose [15] and use it as a feature. However, this method requires the extraction of only clips in which the arms and hands are always in the screen when generating upper body movements, which is labor intensive and may significantly reduce the number of data that can be used. Therefore, in this study, we focus only on head movements and use features of a method to reconstruct 3D facial shapes from images called DECA [16].

2 Listening Motion Generation Model

Figure 1 shows an overview of the proposed model. This model is based on Learning2Listen [7]. It takes the past motion and speaker ID of the dialogue agent as input, as well as the motion, speech features, and speaker ID of the interlocutor, and generates the motion of the dialogue agent as a listener. The features are speech and text, as shown in Fig. 2. The Japanese dialogue data used in this study contains multiple speakers, whereas previous studies used single-speaker data for training. Therefore, it is considered better to input speaker information explicitly. The reason why the speaker IDs are input only for listeners was determined by preliminary experiments.

VQ-VAE is a method that combines variational autoencoder (VAE) and vector quantization (VQ). Including VQ solves the posterior collapse problem with conventional VAE. VQ-VAE is based on the Transformer-based architecture [17] to adapt to time series data. As in Formula (1), where the encoder output $z_e(x)$ is discretized by replacing it with the nearest neighbor code in a set of vectors (codebook), and the discretized latent variable $z_q(x)$ is learned to be input to the decoder for reconstruction.

$$z_q(x) = e_k, \text{ where } k = \arg\min_j \left\| z_e(x) - e_j \right\|^2 \tag{1}$$

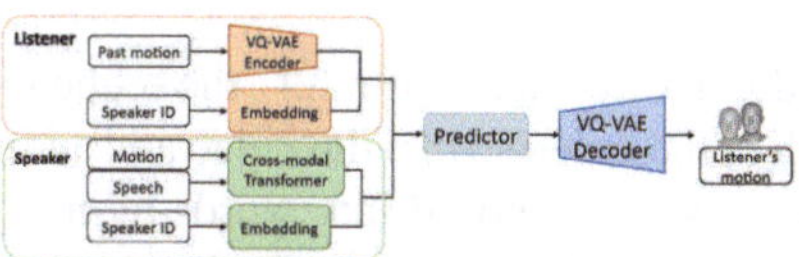

Fig. 1. Model overview

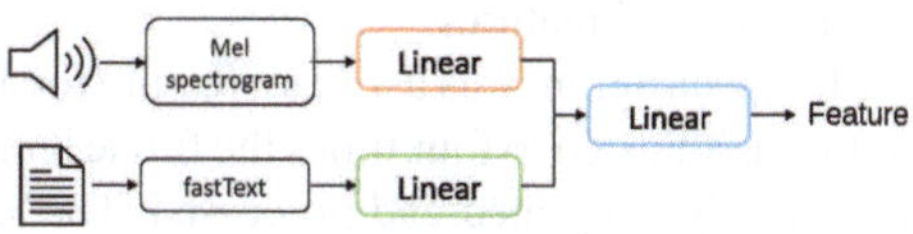

Fig. 2. Speech feature extraction

The advantages of using VQ-VAE for motion generation are that it efficiently preserves human-like behavior, stays within the range of realistic motion, and can predict multinomial distributions for future motion. When generating motion, the code stored in the codebook is used to make predictions, resulting in stable output. This means that the system does not lose its shape or produce unrealistic motion as it continues to generate motion and can output a wide variety of motion.

It receives motion, speech, and text as multimodal input from the speaker. The input from the speaker is not quantized but is integrated using cross-modal attention. Speech and text are converted to features, combined through the all-combining layer, and further converted to speech features by passing through the all-combining layer. The speech features and the speaker's movements are then integrated by the crossmodal Transformer. The speech feature is the query, and the motion is the key and value.

The predictor is Transformer-based and combines the listener features obtained by combining the past motion of the listener discretized by the pre-trained VQVAE encoder and the embedding vector of the speaker ID with the speaker features extracted by the cross-modal Transformer. The cross-modal transformer combines these features and outputs a multinomial distribution for the indices of the VQ-VAE codebook. The codes sampled from the multinomial distribution are then input to the VQ-VAE decoder, which outputs the motion of the interlocutor's speech.

3 Objective Evaluation

3.1 Experimental Data and Features

The model was trained using the Spontaneous Multimodal One-on-one Chat-talk (SMOC) Corpus [18]. This corpus is a large corpus of Japanese multimodal chat dialogues, containing 510 one-on-one chat dialogues between 102 pairs of 107 native Japanese speakers. The data includes audio, video, and transcriptions of the utterances with time information. This experiment used 20 pairs of dialogue data from dialogue numbers D001 to D020, and 21 participants were included in this dialogue.

The frame rate of the video was 30 fps, and then the speech parts were extracted using the timestamps of the transcribed data. Only those with a duration of 64 frames or longer were extracted, and the number of utterances was 5923.

We used DECA [16] for extracting features of head and face, same as L2L. DECA is a method for reconstructing 3D facial shapes by estimating facial posture, facial expression, and shape parameters from facial images according to FLAME 3DMM [19], a parametric face model. In the present study, the concatenation of 50-dimensional facial expression coefficients and 6-dimensional facial postures obtained from this process was used as the head features.

Mel spectrogram was used as a feature for audio. However, as with L2L, the sequence length was adjusted to four times the frame length of the video so that the video and audio can be easily integrated and processed. The text was decomposed into words by using Janome[1] and converted into feature vectors by fastText. We used the fastText model published on huggingface[2]. After generating a vector for the text, we repeated the vector to be the same length as the speech frames.

3.2 Experimental Conditions

We compared two VQ-VAEs: the regular VQ-VAE used in L2L and the VQ-VAE that implements exponential moving average (EMA) and code reset. Exponential moving averages and code resets are learning strategies to improve the performance of VQ-VAE. In both cases, the codebook size was set to 512 codes and 512 dimensions, and the down sampling rate was set to 8. The dialog data was trimmed to 64 frames for training. AdamW was used as the optimization method, with $[\beta_1, \beta_2] = [0.9, 0.98]$. The batch size was set to 64 for regular VQ-VAE and 256 for VQ-VAE with EMA and code reset.

[1] https://github.com/mocobeta/janome.

[2] https://huggingface.co/facebook/fasttext-ja-vectors.

The reason is that code resets are more likely to occur when the batch size is small. The learning rate is based on the scheduling strategy proposed in Transformer [17] and varies with the model dimension and the number of steps. The initial learning rate was set to 2.0. The loss function is as follows, where x is the input, $\hat{x}$ is the reconstructed output, $E(x)$ is the encoder output, and z_q is the discretized code.

$$L = \|x - \hat{x}\| + \|\mathrm{sg}[E(x)] - z_q\|_2^2 + \|\mathrm{sg}[z_q] - E(x)\|_2^2 \tag{2}$$

Here, $\mathrm{sg}[\cdot]$ stands for the stop gradient.

Cross-modal Transformer uses a 2-layer transformer with 512 hidden units and eight heads and Predictor uses a 5-layer transformer with 256 hidden units and eight heads. Adam was used as the optimization method, with $[\beta_1, \beta_2] = [0.9, 0.98]$. The learning rate was initially set to 0.01 using the same scheduling strategy used for VQ-VAE training. The batch size was set to 64, and a cross-entropy loss function was used. The system learned to predict the listener's motion by providing the current eight frames of speaker information in addition to the previous 32 frames of listener and speaker information as input. We compared three conditions: using only speech as the speaker's speech feature, using only text, and using both.

In this study, MSE, FD, and Variation were used as evaluation indices; FD is an index proposed by L2L, and the Frechet distance is calculated directly for the movements predicted by the generative model and the ground-truth movements. The formula for Frechet Distance is as follows, which computes the distribution distance between the ground truth (GT) and the generated motion.

$$FD = \|\mu_{gt} - \mu_{pred}\|^2 + \mathrm{Tr}\left(\Sigma_{gt} + \Sigma_{pred} - 2\sqrt{\Sigma_{gt}\Sigma_{pred}}\right) \tag{3}$$

Variation evaluates the movements' variety by calculating the generated movements' time series variance. Because listeners do not move much, a high Variation is not necessarily good. Since the goal is to generate human-like movements, Variation should be at the same level as the ground truth.

3.3 Results

The results of the objective evaluation of the experiment are shown in Table 1. The overall performance of VQ-VAE was better when EMA and code resets were used, and the usage rate of the codebook was considerably higher, which also results in a higher Variation. When we used the original VQ-VAE, the results for the modality using only speech were higher than those using only text. On the contrary, when using the VQ-VAE with EMA and code reset, multimodal learning performed best, probably due to the increased expressiveness of the codebook and the effect of its increased use. However, the variation values of the VQ-VAEs were still lower than that of the GT, indicating that there is still room for improvement in the prediction part of the behavior.

A visualized example of the generated head motion is shown in Fig. 3. Figure 3 is arranged chronologically from left to right. We used the Python library pyvista [20] for visualization. The inference was performed using 32 frames of the listener's past.

motion and 32 + 8 frames of the speaker's past and present motion and speech, and the motion was generated in an autoregressive fashion.

Table 1. Comparison of performance by VQ-VAE and the modality

VQ-VAE	Modality	MSE	FD(10^3)	Variation	Codebook usage
GT	-	-	-	2.55	-
Original	A	22.19	1.15	1.09	11.7%
	T	24.40	1.27	1.13	11.5%
	A + T	24.23	1.26	1.18	10.9%
EMA&	A	17.35	0.86	1.36	38.3%
Code Reset	T	17.69	0.88	1.43	39.1%
	A + T	17.04	0.85	1.37	39.1%

The generated movements showed that they are generated appropriately to some extent, such as the nodding shown in Fig. 3. In addition, by inputting the speaker ID, the generated motion also changed significantly when the target speaker's motion was large or when the speaker's head tilt was similar to that of the target speaker, and it seems as if differences in motion by the speaker were captured. However, there were many movements in which the same code was selected repeatedly, and the same movements were repeated in some parts. Choosing a code with a little movement made the head movement more unnatural.

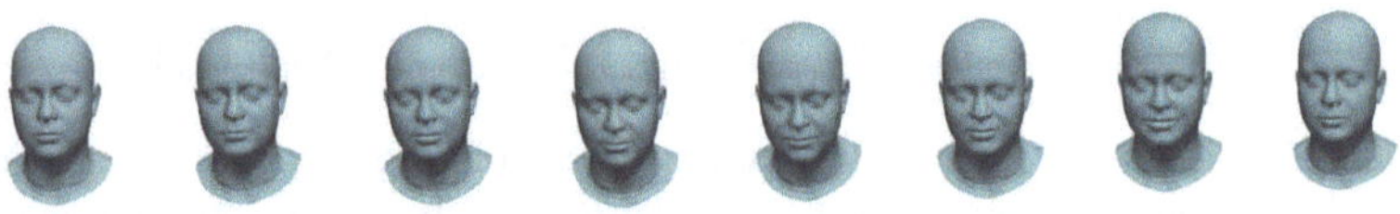

Fig. 3. Visualization of generated head

4 Subjective Evaluation

Next, we evaluated the proposed method by a subjective evaluation. We used the best model in the previous section (speech and text as input and used EMA and code reset) as the proposed method.

4.1 Data Preparation

In this experiment, we compared the actual human motion (GT) and the randomly generated motion (Random) with the proposed method (Proposed). Since the proposed method outputed 32 frames of motion at a time, we appled a window function to overlap 16 frames to avoid discontinuity, and the data was converted to 3D geometry data by DECA, and then visualized by pyvista. The speaker ID used to generate the motion by the proposed method was the same as the speaker ID of the input speech-response pair.

The "Random" condition is the result of decoding a randomly sampled code sequence from the VQ-VAE codebook using EMA and code reset. However, all parameters other than facial expressions and head posture used in DECA's conversion to 3D geometry data are identical. Since the first 32 frames are used as the seed gesture in the generation of the proposed method, the first 32 frames of all videos were processed to stop at the 33rd frame of the image to match the conditions.

The samples for comparison were 20 speech-response pairs randomly selected from the test data from SMOC, and the same samples were used for all comparison conditions. The samples were selected according to the following procedure. First, the samples longer than 6 s were selected. Second, the samples with noise were excluded because a part of the samples in SMOC had masking noise to mask the private information. The selected data were classified into four groups according to the number of frames (i.e., length): 250 frames or less, 250 to 300 frames, 300 to 350.

frames, 350 to 400 frames, and 400 frames or more. Finally, an equal number of samples were extracted from each group. The frame rate was 30 fps. The average length of the selected samples was approximately 8.6 s.

Eleven undergraduate and graduate students (all males) who are native speakers of Japanese participated in the experiment. Each participant evaluated 60 samples (20 samples × 3 conditions). The evaluators rated each head movement on a 5-point scale of 1. Bad, 2. Poor, 3. Fair, 4. Good, and 5.Excellent. Since this experiment aimed to evaluate human-like characteristics, the video was silent and did not include speech sounds. In this way, the evaluation depended only on the motion. The order of presenting the samples was randomized. The response time was unlimited, and the evaluator could view the same sample as many times as they wants.

4.2 Evaluation of Human-Likeliness of the Motion

Figure 4 shows the result of subjective evaluation. The ANOVA did not show significant differences between the three conditions at the significance level of 5%. These results indicate that VQ-VAE with EMA and code resetting can learn human motion accurately and generate human-like motion even when sampled randomly. Next, we examined the samples with high and low scores for each condition. The samples with high scores in the Random condition were basically facing forward and included movements such as nodding and looking up. The tendency was similar in GT, where samples with little or no movement were rated highly, while those with large reactions and those with laughter were rated low. In Proposed, the trend was almost the same, but some samples with low ratings repeated the same movements.

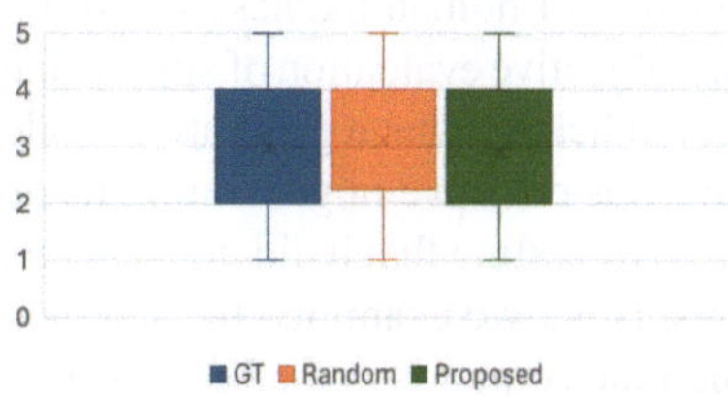

Fig. 4. Subjective evaluation result of human-likeliness of the head motion

4.3 Evaluation of Appropriateness as a Listener

In this experiment, we used the same model used in the previous evaluation experiment and examined the degree to which the generated movements are listener-like by subjective evaluation.

Figure 5 shows the user interface for subjective evaluation. Each video contained a speaker (left) and a listener (right). The listener was a grayscale 3D mesh as before, but the speaker was presented with the actual human motion included in the SMOC along with the speech sound. Two speech-response pairs were presented in the upper and lower rows. The same participants as the previous experiment participated in the experiment. In this experiment, each rater evaluated 40 samples (20 samples $\times$ 2 comparison conditions) in order to compare Proposed–GT pair and Proposed–Random pair. The evaluators compared the two videos and evaluated the difference of appropriateness as a listener in a five-grade scale: -2 as "top is clearly better" to 2 as "bottom is clearly better." The order of presented samples was randomized. We did not limit the response time; the same sample could be viewed multiple times.

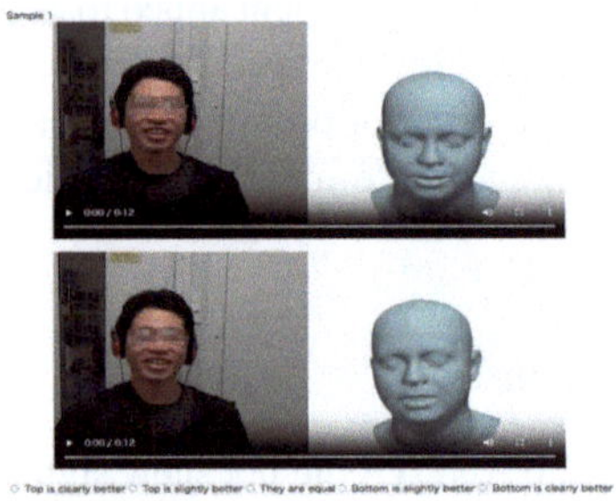

Fig. 5. The user interface of subjective evaluation.

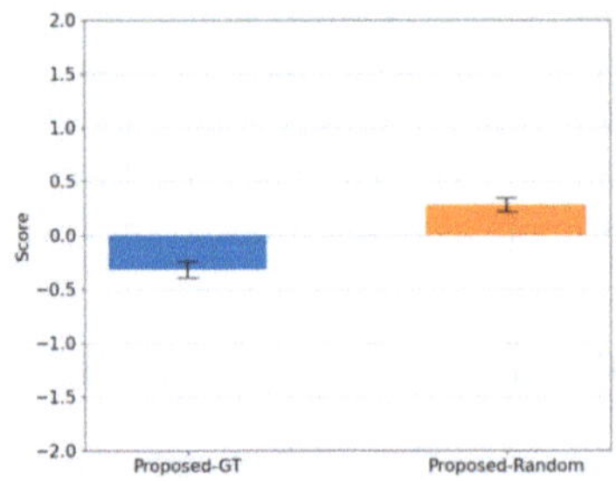

Fig. 6. The evaluation result of the appropriateness as a listener.

Figure 6 shows the results. Here, the positive value means that the Proposed method was preferred, and the negative value means that the other method (GT or Random) was preferred. We conducted the Wilcoxon signed-rank test to test whether the value is different from zero, and we obtained significant differences ($p < 0.001$) for both comparisons.

The comparison between Proposed and Random showed that Proposed was preferred. The subjective evaluation of human likeness showed that Random was as good as the other methods, but the subjective evaluation of appropriateness as a listener showed that Proposed was preferred, indicating that a generative model that considers the interlocutor's speech information was effective in generating listener actions. However, GT was preferred than Proposed, suggesting that it did not generate listening actions as realistic as actual human actions. Next, we examined the samples for which Proposed was preferred and those for which the comparison method was preferred. In the comparison between GT and Proposed, many of the samples for which GT was preferred had more natural timing for nodding and laughing. Conversely, the samples that preferred Proposed were mostly stationary in GT but nodded frequently in Proposed. In the samples where Random were preferred, the timings of laughter in Proposed were unnatural and the Random samples produced few movements.

5 Conclusion

In this study, we investigated a method of motion generation when an anthropomorphic dialogue agent is the listener. Since it is known that the correlation between head movements and speech information is low and between head movements and language information is high in Japanese dialogue, we tested whether the model performance can be enhanced by inputting text information. The experimental results showed that speech information is more important than text information in generating listening behavior but that multimodal learning is effective.

In the future, we aim to improve performance by improving the prediction part of the generative model.

Acknowledgments. Part of this work was supported by JSPS Kakenhi JP23K20725 and JP21H00895.

References

1. Cassell, J., Sullivan, J., Prevost, S., Churchill, E.: Embodied Conversational Agents. MIT Press (2000)
2. Cassell, J., Vilhj´almsson, H.H., Bickmore, T.: Beat: the behavior expression animation toolkit. In: Proceedings of the 28th Annual Conference on Computer Graphics and Interactive Techniques, pp. 477–486 (2001)
3. Salvi, G., Beskow, J., Al Moubayed, S., Granstr¨om, B.: Synface—speech-driven facial animation for virtual speech-reading support. EURASIP J. Audio Speech Music Process. **2009**, 1–10 (2009)
4. Deichler, A., Mehta, S., Alexanderson, S., Beskow, J.: Diffusion-based co-speech gesture generation using joint text and audio representation. In: GENEA: Generation and Evaluation of Non-verbal Behaviour for Embodied Agents Challenge 2023 (2023)
5. Jonell, P., Kucherenko, T., Henter, G.E., Beskow, J.: Let's face it: probabilistic multi-modal interlocutor-aware generation of facial gestures in dyadic settings. In: Proceedings of the 20th ACM International Conference on Intelligent Virtual Agents, pp. 1–8 (2020)
6. Zhou, M., Bai, Y., Zhang, W., Yao, T., Zhao, T., Mei, T.: Responsive listening head generation: a benchmark dataset and baseline (2022). https://arxiv.org/abs/2112.13548
7. Ng, E., et al.: Learning to listen: Modeling non-deterministic dyadic facial motion. In: Proceedings of the IEEE/CVF Conference on Computer Vision and Pattern Recognition, pp. 20395–20405 (2022)
8. Song, L., Yin, G., Jin, Z., Dong, X., Xu, C.: Emotional listener portrait: Realistic listener motion simulation in conversation. In: 2023 IEEE/CVF International Conference on Computer Vision (ICCV), pp. 20782–20792 (2023). https://doi.org/10.1109/ICCV51070.2023.01905
9. Ng, E., Subramanian, S., Klein, D., Kanazawa, A., Darrell, T., Ginosar, S.: Can language models learn to listen? In: 2023 IEEE/CVF International Conference on Computer Vision (ICCV), pp. 10049–10059 (2023). https://doi.org/10.1109/ICCV51070.2023.00925

10. Wu, B., Liu, C., Ishi, C.T., Ishiguro, H.: Modeling the conditional distribution of co-speech upper body gesture jointly using conditional-gan and unrolled-gan. Electronics **10**(3), 228 (2021)
11. Takeuchi, K., Hasegawa, D., Shirakawa, S., Kaneko, N., Sakuta, H., Sumi, K.: Speech-to-gesture generation: a challenge in deep learning approach with bidirectional lstm. In: Proceedings of the 5th International Conference on Human Agent Interaction, pp. 365–369 (2017)
12. Wu, B., Liu, C., Ishi, C.T., Ishiguro, H.: Probabilistic human-like gesture synthesis from speech using GRU-based WGAN. In: Companion Publication of the 2021 International Conference on Multimodal Interaction, pp. 194–201 (2021)
13. Ishii, R., Katayama, T., Higashinaka, R., Tomita, J.: Generating body motions using spoken language in dialogue. In: Proceedings of the 18th International Conference on Intelligent Virtual Agents, pp. 87–92 (2018)
14. Ishi, C.T., Haas, J., Wilbers, F.P., Ishiguro, H., Hagita, N.: Analysis of head motions and speech, and head motion control in an Android. In: 2007 IEEE/RSJ International Conference on Intelligent Robots and Systems, pp. 548–553. IEEE (2007)
15. Cao, Z., Hidalgo Martinez, G., Simon, T., Wei, S., Sheikh, Y.A.: Openpose: realtime multi-person 2D pose estimation using part affinity fields. IEEE Trans. Pattern Anal. Mach. Intell. (2019)
16. Feng, Y., Feng, H., Black, M.J., Bolkart, T.: Learning an animatable detailed 3D face model from in-the-wild images. ACM Trans. Graph. (ToG), Proc. SIGGRAPH **40**(4), 88–18813 (2021)
17. Vaswani, A., et al.: Attention is all you need. In: Proceedings of the 31st International Conference on Neural Information Processing Systems, pp. 6000–6010 (2017)
18. Yamazaki, Y., Chiba, Y., Nose, T., Ito, A.: SMOC corpus: a large-scale Japanese spontaneous multimodal one-on-one chat-talk corpus for dialog systems. Acoust. Sci. Technol. **42**(4), 210–213 (2021)
19. Li, T., Bolkart, T., Black, M.J., Li, H., Romero, J.: Learning a model of facial shape and expression from 4D scans. ACM Trans. Graph. **36**(6), 194–201 (2017)
20. Sullivan, C.B., Kaszynski, A.: PyVista: 3D plotting and mesh analysis through a streamlined interface for the visualization toolkit (VTK). J. Open Source Softw. **4**(37), 1450 (2019). https://doi.org/10.21105/joss.01450

MS-HANet: Multi-scale Hypergraph Attention Network for EEG-Based Emotion Recognition

Zaijun Wang[1(✉)], Ruizhe Yang[1], Yuheng Jiang[1], and Yeteng Yan[2]

[1] Research Base for Civil Aviation Flight Technology and Flight Safety, Civil Aviation Flight University of China, Guanghan, China
tiantian20030315@126.com

[2] Research Base for Air Traffic Management, Civil Aviation Flight University of China, Guanghan, China

Abstract. In recent years, emotion recognition based on electroencephalogram (EEG) has garnered increasing attention due to its ability to capture subtle physiological signals beyond facial or voice cues. However, traditional models often struggle with incomplete feature extraction and inadequate fusion strategies. In this study, a novel framework named Dynamic Hypergraph-based Gated Feature fusion with ECA attention (DHGF-ECA Network) is proposed. This framework combines dynamic hyperedge construction, multi-scale feature extraction, and an innovative feature fusion mechanism. First, we exploit the time-varying characteristics of phase locking value (PLV) to build dynamic functional connectivity, and incorporate an attention mechanism to adaptively adjust hyperedge weights. Second, multi-scale feature extraction is utilized to capture both local and global EEG signal patterns. Third, we introduce time-varying nonlinear entropy (TVNE) as the principal feature, while differential entropy (DE) and the Hurst exponent serve as auxiliary features. A gated feature selection strategy then fuses these features, enhancing discriminative capability. This method achieved an average high accuracy of 96.3% in the three-class classification task on the SEED dataset, outperforming other existing methods.

Keywords: EEG emotion recognition · Dynamic hypergraph · style · Multi-scale feature extraction · Gated feature fusion

1 Introduction

Electroencephalography (EEG) serves as a critical modality for emotion recognition due to its ability to capture genuine neural activity beyond overt behaviors like facial expressions [1]. Early research focused on handcrafted features extracted from time, frequency, and nonlinear domains, with differential entropy (DE) emerging as a particularly effective metric for capturing amplitude and frequency information in short EEG segments. Notably, Zheng and Lu [2] demonstrated significant performance gains on the SEED dataset by integrating DE into deep belief networks, establishing a standard benchmark for EEG-based affective computing. More recent efforts incorporate deep learning to automatically learn feature representations, further improving classification accuracy and robustness.

A. Ito (Ed.): ICMSSP 2025, LNNS 1637, pp. 113–123, 2025.
https://doi.org/10.1007/978-3-032-05994-9_11

However, emotions involve distributed neural processes spanning multiple brain areas, necessitating the modeling of functional connectivity (FC) through graph-based techniques. Song et al. [3] introduced a Dynamical Graph Convolutional Neural Network (DGCNN) that adaptively learns adjacency relations, achieving state-of-the-art results on SEED. Subsequently, attention-based Graph Neural Networks (GNNs) have demonstrated remarkable accuracies, yet they often capture only pairwise electrode relationships. Meanwhile, hypergraph learning has attracted growing interest by connecting multiple EEG channels in a single hyperedge, offering a richer way to represent higher-order interactions [5]. Moreover, recent work in EEG-based emotion recognition suggests that no single feature type suffices to capture the complexity of brain activity, highlighting the importance of multi-scale and adaptive fusion strategies [6].

Motivated by these gaps, this paper proposes a novel EEG emotion recognition framework that unifies dynamic hypergraph connectivity modeling and entropy-based multi-scale feature extraction. Specifically, time-varying phase-locking value (PLV) is used to construct hyperedges, and an attention mechanism further refines hyperedge weights. We integrate Time-Varying Network Entropy (TVNE), DE, and the Hurst exponent to capture complementary information, fusing them through a gated mechanism that adaptively emphasizes the most discriminative features. Our experiments on the SEED dataset illustrate that this approach yields competitive performance, underscoring the utility of dynamic hypergraphs and multi-scale descriptors in EEG-based affective computing.

2 Related Work

2.1 EEG Feature Extraction

Handcrafted EEG features traditionally include time-domain statistics and frequency-domain powers. Among these, DE remains prominent for its capacity to compactly characterize signal distributions across multiple bands [2]. On the SEED dataset, Zheng and Lu's DE-based method exceeded 80% accuracy, outperforming simpler spectral-power-only approaches [2]. Other works have investigated nonlinear metrics such as Hurst exponent and fractal dimension to capture long-term dependencies, showing accuracy boosts when combined with classical descriptors [9]. Hybrid approaches that feed handcrafted features into neural networks have also proven fruitful. For instance, Han et al. incorporated DE feature maps into a multi-scale CNN pipeline and reported accuracy improvements on SEED [7]. Overall, research underscores the complementarity of spectral and complexity-based features for robust emotion classification.

2.2 Functional Connectivity and Graph-Based Modeling

Emotion-related EEG signals frequently manifest as correlated activity across multiple brain regions [1], spurring widespread adoption of functional connectivity measures like PLV, coherence, or correlation to construct brain networks [4]. Song et al. [3] were among the first to embed these connectivity patterns in a deep learning framework, introducing a DGCNN that adaptively refines electrode adjacency. This method surpassed conventional CNNs by capturing dynamic inter-channel interactions. Further innovations include integrating attention mechanisms into Graph Convolutional Networks, as illustrated by Chen et al. [8], who achieved 99.4% accuracy on SEED by jointly attending to critical electrodes and frequency components. Although graph-based models provide a powerful depiction of pairwise relationships, they may overlook complex groupwise dependencies, paving the way for hypergraph-based approaches [5].

2.3 Hypergraph Learning in EEG Emotion Recognition

Hypergraphs extend standard graphs by allowing edges (hyperedges) to connect multiple nodes at once, thus modeling group-level brain interactions during emotional processing [5]. Li et al. employed a spatial–temporal hypergraph convolutional network (STHGCN), showing state-of-the-art performance on SEED by capturing multi-channel correlations more effectively [5]. In a related line, Pan et al. [6] built an online multi-hypergraph fusion method for cross-subject emotion recognition, updating hypergraph structures in real time to reflect changing neural states. Ji et al. [10] developed FC-HAT to model high-order brain connectivity via hypergraph attention. These results suggest that hypergraph representations can address the higher-order connectivity overlooked in conventional pairwise models.

2.4 Multi-scale Feature Extraction and Adaptive Fusion

EEG signals contain rich patterns across temporal, spectral, and spatial scales [7]. Consequently, numerous works employ multi-scale filters or wavelet transforms to extract complementary features. Cheng *et al.* [11] combined a multi-scale dynamic CNN with a gated transformer, adaptively highlighting crucial time segments across scales, while Wang *et al.* [12] leveraged multi-scale convolution fused via an attention module for improved classification. Gated fusion mechanisms further refine these multi-scale features by dynamically adjusting their relative contributions in accordance with the underlying emotional context [6]. Such adaptivity is crucial, as certain emotional states may exhibit stronger connectivity signatures, whereas others display more distinctive spectral or fractal characteristics.

Integrating hypergraph-based connectivity with an adaptive fusion of entropy features remains a promising yet less explored direction. Existing hypergraph models typically rely on a single feature set (e.g., DE alone), whereas multi-scale entropy features like *TVNE*, *Hurst exponent*, and *DE* could reveal additional information about the complex neural underpinnings of emotion. A gated selection strategy can further isolate the most pertinent features, enhancing interpretability and predictive power. This combined perspective—hypergraph connectivity plus multi-scale descriptors—motivates our proposed approach, which dynamically constructs and refines hyperedges while extracting diverse entropy features and selectively merging them via gating.

3 Background Theory and Method

This chapter provides a comprehensive description of our EEG-based emotion recognition approach, encompassing feature extraction and fusion, as well as the overall model construction.

3.1 Feature Extraction

Differential Entropy (DE) is frequently used to capture spectral power variations across multiple frequency bands ($\delta, \theta, \alpha, \beta, \gamma$). For a Gaussian process with variance σ^2, the differential entropy can be written as

$$h(X) = \frac{1}{2} ln\left(2\pi e\sigma^2\right) \tag{1}$$

In practice, σ^2 is the variance of a short EEG segment. A higher variance leads to a larger DE value.

Hurst exponent (H) estimates long-range dependence or persistence in time series. Emotional states may modify the EEG's long-memory patterns. Hence, H can reveal how persistent (or anti-persistent) the brain's electrical activity is under different affective conditions. Using the classic R/S analysis: Compute the range R of the cumulative deviation from the mean, and the standard deviation S. The ratio R/S scales with segment size n as

$$\frac{R}{S} \propto n^H. \tag{2}$$

EEG signals exhibit nonlinear, non-stationary characteristics. Time-Varying Nonlinear Entropy (TVNE) reflects instantaneous complexity or irregularity over sliding windows. Emotional stimuli can induce significant changes in such complexity, offering a complementary perspective for emotion classification. Given a time series $\{x_1, \ldots, x_N\}$, an embedding dimension m, and a similarity threshold r:

Form length-m templates $\mathbf{u}_i^m = (x_i, \ldots, x_{i+m-1})$.

Determine their matching counts within distance r.

Let:

B^m = average matching probability for length- m templates,

A^m = average matching probability for length-$(m+1)$ templates.

The sample entropy is:

$$SampEn(m, r) = -ln\left(\frac{A^m}{B^m}\right) \tag{3}$$

3.2 Feature Concatenation and Fusion Mechanism

We apply band-pass filtering to split the EEG data into five frequency bands, namely δ (1–4 Hz), θ (4–8 Hz), α (8–13 Hz), β (13–30 Hz), and γ (30–50 Hz). For each band and each of the 62 channels, we compute the above three features.

The combination of 62 channels and 5 bands results in 310 coefficients. By selecting three specific features, we obtain 930 coefficients per trial. This leads to a 930-dimensional feature vector for each trial.

This approach offers a richer input for the downstream gating and attention mechanisms, as well as the hypergraph model.

3.3 Dynamic Hypergraph Construction

Emotion-related EEG signals often exhibit distinct spatiotemporal dynamics in terms of functional connectivity across different brain regions. We employ a dynamic hypergraph to represent such interactions over time. A hypergraph allows each "hyperedge" to simultaneously connect multiple nodes (channels), which can more naturally describe higher-order neuronal interactions.

Phase Locking Value (PLV)

For each EEG channel, within a chosen time window of length T, we apply the Hilbert transform to obtain the instantaneous phase:

$$\text{phase}_i(t) = \angle(\text{H}\{x_i(t)\}) \tag{4}$$

where $x_i(t)$ is the EEG signal of channel I, H[·] denotes the Hilbert transform, and $\angle(\cdot)$ represents taking the argument (phase) of a complex number.

For any two channels i and j, we compute the **phase locking value** PLV_{ij} based on the phase difference $\text{phase}_i(t) - \text{phase}_j(t)$:

$$\text{PLV}_{ij} = \left| \frac{1}{T} \sum_{t=1}^{T} e^{\text{i}\left[\text{phase}_i(t) - \text{phase}_j(t)\right]} \right| \tag{5}$$

PLV_{ij} ranges from 0 to 1, where a higher value indicates stronger phase synchronization between the two channels within that window.

As the window slides forward in time, each $\text{PLV}^{(w)}$ reflects the functional connectivity of a different temporal segment. Stacking these matrices yields a time-varying connectivity profile, where w indexes the w-th time window.

Hyperedge Extraction Strategy

For each time window w, after obtaining $\text{PLV}_{ij}^{(w)}$, we typically set a threshold θ (or select a high percentile of PLV values) to emphasize significant connections. This yields a binary adjacency matrix $A_{ij}^{(w)}$, as Fig. 1 shown

$$A_{ij}^{(w)} = \begin{cases} 1, \text{ if } \text{PLV}_{ij}^{(w)} \geq \theta, \\ 0, \text{ otherwise.} \end{cases} \tag{6}$$

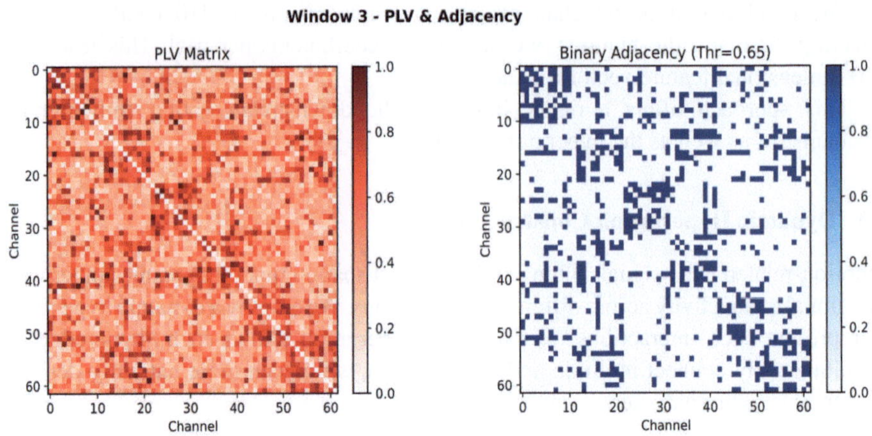

Fig. 1. PLV Matrix and Binary Adjacency Matrix in 3rd Window

In the resulting adjacency graph $G(w)$, each connected component containing multiple nodes (channels) can be treated as a single "hyperedge." Since a hyperedge may link more than two nodes at once, it better captures group-wise synchronization or higher-order interactions.

Incidence Matrix

Let $\mathrm{E}^{(w)} = \{e_1, e_2, \ldots, e_M\}$ be the hyperedge set in the w-th window. We define an incidence matrix $\boldsymbol{H}(w)$ of size $(N \times M)$, where N is the number of channels (e.g., 62).

If channel i belongs to hyperedge e_j, we set $\boldsymbol{H}_{i,j}^{(w)} = 1$, otherwise 0. Such a matrix makes it convenient to incorporate hyperedge structure into subsequent model architectures.

3.4 Construction and Training

We designate our proposed model as the Dynamic Hypergraph-based Gated Feature fusion with ECA attention (DHGF-ECA Network). Building on pre-extracted EEG features, our framework is designed to integrate dynamic connectivity structures with multiscale descriptors in a unified pipeline. As shown in Fig. 2, the model comprises four primary components after feature extraction: (1) Feature Integration Module, (2) Residual Network, (3) Bi-LSTM Stack, and (4) Classification Head.

ECA Attention

We employ an Efficient Channel Attention (ECA) mechanism to assign channel-wise importance factors. Let $\mathbf{F} \in \mathbb{R}^{C \times D}$ represent the concatenated features across C frequency sub-bands and D feature dimensions. By aggregating global statistics and applying a 1D convolution followed by a sigmoid, ECA yields a vector of attention coefficients α. The features are then scaled:

$$\mathbf{F}_{\mathrm{ECA}} = \alpha \odot F \tag{7}$$

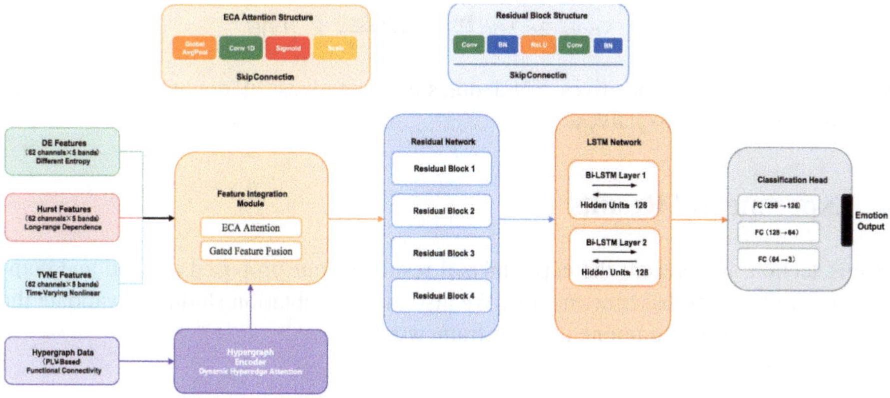

Fig. 2. The architecture of the DHGF-ECA Network

where $\odot$ denotes *elementwise* multiplication along the channel dimension.

Gated Feature Fusion

We further *introduce* a gating mechanism to adaptively blend distinct feature streams (e.g., DE, Hurst, TVNE, and hypergraph representations). Let **G** denote the concatenated multi-source feature tensor, and **W**, **b** be learnable parameters. A sigmoidal gate σ modulates each dimension:

$$\mathbf{G}_{\text{fused}} = \sigma(\mathbf{WG} + \mathbf{b}) \odot G \tag{8}$$

Residual Network and Bi-LSTM Stack

The integrated features $\mathbf{F}_{\text{ECA}}$ and $\mathbf{G}_{\text{fused}}$ are then forwarded to a Residual Network composed of four stacked residual blocks. Each block contains two convolutional layers (with Batch Normalization and ReLU), plus a skip connection that mitigates vanishing gradients and stabilizes training. The resulting higher-level representations are fed into a two-layer Bi-LSTM module, each Bi-LSTM layer having 128 hidden units in the forward and backward directions. This recurrent structure further captures temporal dependencies inherent in EEG signals, allowing the model to learn subtle changes over time.

Evaluation Metrics

To evaluate the performance of the DHGF-ECA Network model, four standard classification metrics were employed: accuracy, precision, recall, and F1 score. The formulas are as follows:

$$Accuracy = \frac{TP+TN}{TP+TN+FP+FN} \times 100\% \tag{9}$$

$$Precision = \frac{TP}{TP+FP} \times 100\% \tag{10}$$

$$Recall = \frac{TP}{TP+FN} \times 100\% \tag{11}$$

$$F1\ Score = 2 \times \frac{\text{Precision} \times \text{Recall}}{\text{Precision} + \text{Recall}} \times 100\% \tag{12}$$

Where TP denotes true positives, TN denotes true negatives, FP denotes false positives and FN denotes false negatives.

4 Experimental Results

In this section, we present the experimental results of our proposed method and compare it with Existing baseline models. We also conduct ablation studies to validate the effectiveness of each component in our framework.

4.1 Dataset

The open-source SEED dataset contains 62-channel EEG recordings from 15 subjects, labeled for three emotions (positive, neutral, negative). Data was sampled at 200Hz with standard pre-processing; standard splits were adopted [2].

4.2 Overall Performance

Our proposed multi-scale hypergraph neural network with attention mechanism achieved an average accuracy of 96.3% (±0.42%) on the SEED dataset, demonstrating its effectiveness in EEG-based emotion recognition. Table 1 presents the detailed performance metrics across different emotion categories. Figure 3 shows the accuracy/loss curves during the training process.

Table 1. Performance metrics on the SEED dataset

Emotive Category	Accuracy	F1-Score	Precision	Recall
Positive	97.2 ± 0.38	97.0 ± 0.41	97.3 ± 0.36	96.8 ± 0.43
Neutral	95.8 ± 0.51	95.6 ± 0.48	95.9 ± 0.45	95.4 ± 0.52
Negative	95.9 ± 0.47	96.0 ± 0.45	96.1 ± 0.42	96.0 ± 0.48
Average	96.3 ± 0.42	96.2 ± 0.44	96.4 ± 0.39	96.1 ± 0.46

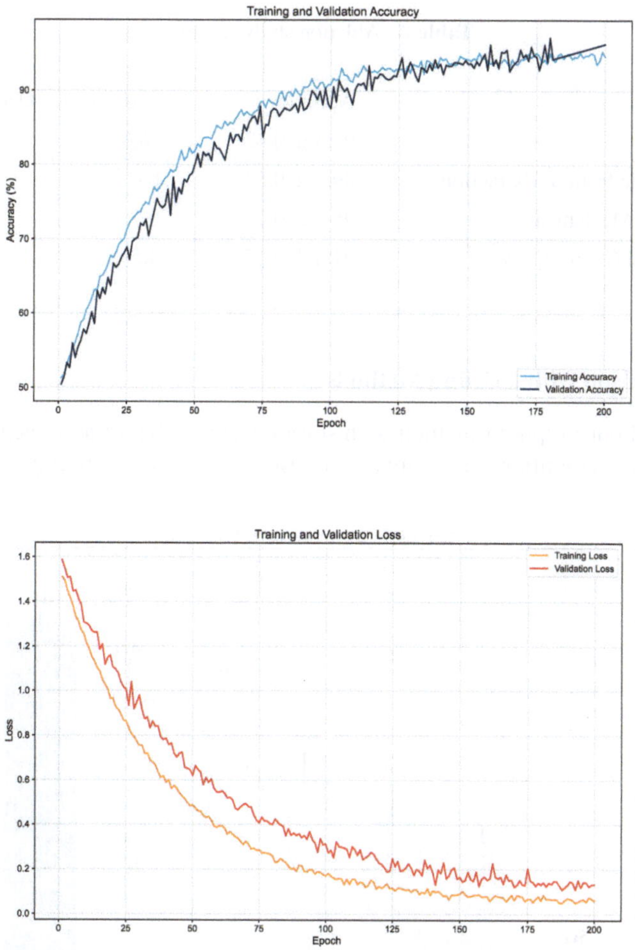

Fig. 3. Accuracy/Loss curves

4.3 Ablation Study

To evaluate the contribution of each component in our proposed framework, we conducted a comprehensive ablation study by removing one component at a time and measuring the resulting performance. The results are summarized in Table 2.

Ablation studies reveal that all components play important roles in the model. Functional connectivity is crucial for accurate emotion recognition. Hypergraph learning can capture unique relationships. Multi-scale feature extraction enhances the model's recognition ability, and the attention mechanism helps the model focus on key elements.

Table 2. Ablation study results

Model Variant	Accuracy	Performance Decrease
Full Model	96.3 ± 0.42	N/A
w/o Multi-scale Feature Extraction	89.7 ± 0.53	6.6
w/o Attention Mechanism	96.0 ± 0.78	0.3
w/o Functional Connectivity	91.8 ± 0.57	4.5

4.4 Comparison with Existing Methods

We compared our proposed method with several state-of-the-art approaches for EEG-based emotion recognition. The comparison results are presented in Fig. 4.

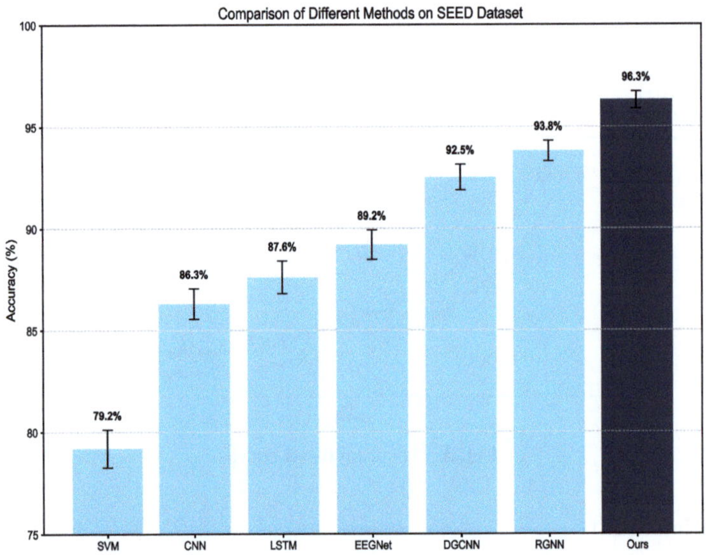

Fig. 4. Comparison of Different Methods

Our proposed method significantly outperforms existing approaches across all evaluation metrics.

5 Summary

In this task, We presents a multi-scale hypergraph neural network with attention mechanism for EEG-based emotion recognition. The proposed framework integrates functional connectivity analysis with hypergraph learning to capture complex brain region interactions during emotional processing. Experiments on the SEED dataset demonstrate superior performance with 96.3% accuracy, significantly outperforming existing methods. Multi-scale feature extraction combined with attention mechanisms enables the model to focus on the most relevant neural patterns. Despite these advances, individual differences in emotional responses remain challenging to address. Future work should explore personalized models and investigate the framework's generalizability across different EEG datasets with varying recording conditions.

Acknowledgment. I am deeply indebted to my advisor Zaijun Wang and would like to convey my heartfelt gratitude. Throughout the entire research process, her guidance and support have been truly invaluable.

This research is supported by Key Laboratory of Flight Techniques and Flight Safety, CAAC (FZ2022ZZ02); Engineering Technology Research Center for Civil Aviation Flight Technology and Flight Safety Project (FZ2022ZZ02); College Students' Innovation and Entrepreneurship Training Program (202410624019).

References

1. Jenke, R., Peer, A., Buss, M.: Feature extraction and selection for emotion recognition from EEG. IEEE Trans. Affect. Comput. **5**(3), 327–339 (2014)
2. Zheng, W.-L., Lu, B.-L.: Investigating critical frequency bands and channels for EEG-based emotion recognition with deep belief networks. IEEE Trans. Affect. Comput. **7**(3), 162–175 (2016)
3. Song, T., Zheng, W., Song, P., Cui, Z.: EEG emotion recognition using dynamical graph convolutional neural networks. IEEE Trans. Affect. Comput. **11**(3), 532–541 (2020)
4. Zheng, F., Hu, B., Zheng, X., Ji, C., Bian, J., Yu, X.: Dynamic differential entropy and brain connectivity features based EEG emotion recognition. Int. J. Intell. Syst. **37**(12), 12511–12533 (2022)
5. Li, M., Qiu, M., Zhu, L., Kong, W.: Feature hypergraph representation learning on spatial–temporal correlations for EEG emotion recognition. Cogn. Neurodyn. **17**(5), 1271–1281 (2023)
6. Pan, T., Ye, Y., Zhang, Y., Xiao, K., et al.: Online multi-hypergraph fusion learning for cross-subject emotion recognition. Inf. Fus. **108**, 102338 (2024)
7. Han, L., Zhang, X., Yin, J.: EEG emotion recognition based on the TimesNet fusion model. Appl. Soft Comput. **159**, 111635 (2024)
8. Chen, W., Liao, Y., Dai, R., Dong, Y., Huang, L.: EEG-based emotion recognition using graph convolutional neural network with dual attention mechanism. Front. Comput. Neurosci. **18**, 1416494 (2024)
9. Yousefi, M.R., Dehghani, A., Taghaavifar, H.: Enhancing the accuracy of electroencephalogram-based emotion recognition through long short-term memory recurrent deep neural networks. Front. Hum. Neurosci. **17**, 1174104 (2023)
10. Ji, J., Ren, Y., Lei, M.: FC-HAT: hypergraph attention network for functional brain network classification. Inf. Sci. **599**, 1301–1316 (2022)
11. Cheng, Z., Bu, X., Wang, Q., Yang, T., Tu, J.: EEG-based emotion recognition using multi-scale dynamic CNN and gated transformer. Sci. Rep. **14**(1), 31319 (2024)
12. Wang, Y., Zhang, X., Jin, X., et al.: EEG emotion recognition by fusion of multi-scale features. In: Proceedings of the 1st International Conference on Data Science and Engineering (ICDSE), pp. 85–92. IEEE (2024)

Visual Differential Signal Processing for Generative Image Editing

Peng Wei, Yuanting Zhang, Quanwei Wu, and Yi Wang(✉)

School of Computer Science and Technology, Dongguan University of Technology, #1 University Road, Dongguan 523808, Guangdong, China
{221115238,231215020,wangyi}@dgut.edu.cn

Abstract. Deep generative models, particularly Generative Adversarial Networks (GANs), have made significant strides in the fields of image generation and manipulation. However, current image processing methods face two critical challenges: insufficient control over the outputs of generative models and a lack of interpretability regarding changes in the latent space of these models. To address these challenges, we propose a novel method called Fredit to facilitate controllable image editing in the frequency space. The core idea is to design a learnable Finite-Impulse-Response (FIR) filter to model and manipulate specific frequency components with respect to the semantic changes of the input image data. Since our DSP components are designed for vision tasks and exploit the attention mechanism as in visual transformers, we call the proposed module Visual DSP (VDSP) and propose a contrastive representation loss to supervise the learning of our VDSP filter. We extensively evaluate the Fredit method on two large-scale datasets for image editing tasks. Through comprehensive experiments, we demonstrate that Fredit can significantly enhance the controllability and interpretability of generative image editing, effectively allowing for the manipulation of semantic changes in images.

Keywords: Image attribute editing · GAN · Frequency filter · Digital signal processing

1 Introduction

Deep generative models have revolutionized fields like image synthesis, text generation, and data augmentation[1]. These models are designed to learn data distributions and generate new, realistic data samples that mimic the underlying patterns of real-world data. In particular, generative adversarial networks (GANs) [2] have significantly advanced the field of image generation and manipulation. They have found widespread applications such as facial image manipulation, photo enhancement, and artistic image generation.

A. Ito (Ed.): ICMSSP 2025, LNNS 1637, pp. 124–135, 2025.
https://doi.org/10.1007/978-3-032-05994-9_12

Despite their successes, several challenges remain with GANs. One key limitation is controllable deep data generation, i.e., the ability to precisely control generated outputs in terms of specific attributes or features [3]. While some progress has been made, such as the manipulation of facial expressions or object attributes in images [4], the latent space of many generative models is often highly complex and poorly understood, making it challenging to direct the generation process reliably.

Another key limitation in attribute editing for image synthesis is interpretability. Users often want to modify specific image attributes without inadvertently altering unrelated features. However, precise and intuitive editing is challenging due to the entangled nature of latent representations. Current models lack transparent mechanisms to predict how changes in latent space affect individual attributes, which can result in undesirable side effects.

On the other hand, it is well known that in signal processing certain signal characteristics are more easily analyzed in the frequency domain, whereas they are difficult to disentangle and process in the spatial domain. In this paper, we propose to investigate the design of controllable deep data generation under targeted properties in the frequency domain for the downstream task of image editing. Specifically, we integrate into the generative model (e.g., StyleGAN) a learnable encoder-and-decoder structure based on Differentiable Digital Signal Processing (DDSP) module [5] to achieve more precise and realistic editing of generated image data with desired properties.

The main contributions of this paper are:

- We demonstrate that DDSP can be adapted to use for image editing tasks. Specifically, we design and implement an LTI-FIR filter with a vision transformer (ViT) structure, namely VDSP, and incorporated it into the latent space of StyleGAN generator.
- We propose a contrastive representation loss to supervise the learning of VDSP filter by aligning relevant frequency components with referenced image pairs with modification of the targeted attribute.
- We show that our approach based on VDSP filter can significantly improve the data quality, controllability and interpretability of image generation comparing with existing methods.

2 Related Work

Several GAN-based models have designed controllable data generation for image synthesis [3]. For example, Conditional GANs (cGAN) allow for some level of control over the generated outputs by conditioning on labels or other variables [6]. However, such conditional generation by itself is difficult to achieve fine-grained control (e.g., adjusting specific attributes like age and gender). To address this issue, StyleGAN introduced a style-based generator that operates in an intermediate latent space [4]. This allows for more control over the generated image by adjusting "style" at different levels of detail, from coarse features (e.g., face shape) to fine details (e.g., skin texture).

However, precise control over more complex attributes remains a challenge. In particular, fine-grained, targeted edits are often not intuitive, and certain features can be difficult to modify independently. For instance, StyleGAN struggles with generating images in edge cases, such as unusual facial features, extreme lighting conditions, or non-standard objects. The lack of interpretablity also complicates the control of how certain edits manifest in generated images, leading to biases in semantic understanding of the latent space. There are several variants of StyleGAN proposed to improve image editing. For example, InterfaceGAN [7] trains a Support Vector Machine (SVM) to learn the decision boundary hyperplane for each binary attribute. StyleRes [8] introduces a novel image inversion framework and training process to achieve high-fidelity image inversion and high-quality attribute editing. By learning residual features in higher-dimensional latent encodings and transforming them, it can adapt to manipulation within the latent space.

To leverage frequency information, Cai et al. [9] proposed a framework for joint optimization of pixel space and Fourier spectrum—Frequency Domain Image Transformation (FDIT) to enhance the image generation process. However, its over-reliance on high-frequency information may, in some cases, result in insufficient capture and expression of complex style changes. Shi et al. [3] introduced a GAN-based method that uses local generators to provide explicit spatial control over the synthesized output. Fourier features are incorporated for positional encoding, enabling the generator to better capture the details and structure of the image during synthesis. However, this approach requires semantic segmentation masks as training data, limiting the model's applicability on unlabeled datasets. Engel et al. [5] combined the interpretable structure of classic digital signal processing (DSP) elements (e.g., filters, oscillators, reverberation) with the flexibility and power of deep learning, and introduced the Differentiable DSP (DDSP) library for tasks like singing voice conversion, music synthesis, and speech enhancement[1]. As focusing on neural audio synthesis, the key components include oscillator models, envelope generators, and filters which are all differentiable with respect to parameters of the neural network. This enables end-to-end training of models to learn from raw data while retaining the structure and interpretability of audio signal processing.

[1] https://magenta.tensorflow.org/ddsp.

3 Methodology

Inspired by DDSP, we propose to develop a more interpretable and controllable approach for image attribute editing by introducing data-driven DSP components to modern image generative models such as StyleGAN. Since our DSP components are designed for vision tasks and exploit the attention mechanism as in visual transformers (ViT) [10], we call the proposed module Visual DSP (VDSP). The core idea is to design a learnable Finite-Impulse-Response (FIR) filter to model and manipulate specific frequency components with respect to the semantic changes of the input image data. The following sections elaborate on the design in detail with main components of VDSP illustrated in Fig. 1.

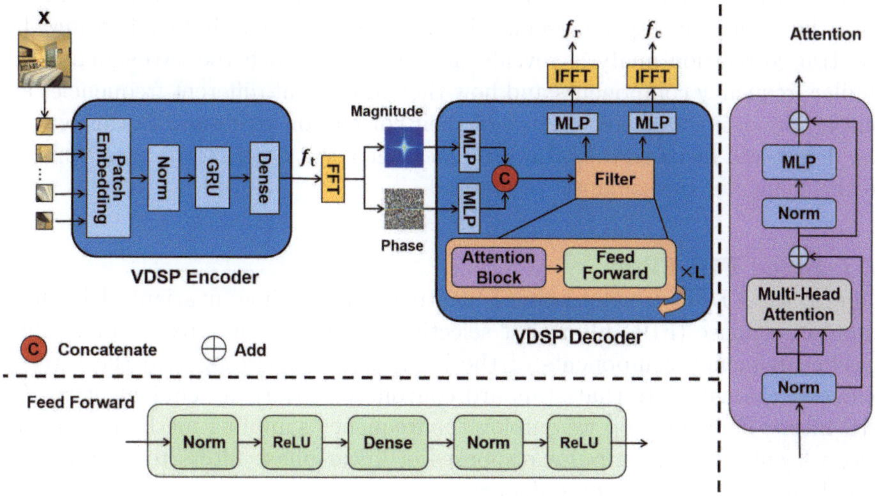

Fig. 1. The main components of the proposed VDSP module for latent feature editing

3.1 Visual DSP (VDSP) Encoder

The VDSP encoder decomposes the input image from the spatial domain into the frequency domain to extract frequency-domain features. As typically done with ViTs, we first divide the input image into non-overlapping patches each of 16×16 pixels. The image data are then flattened and linearly projected to create *patch embedding* sequence representations with added positional encodings. This allows the VDSP encoder to effectively capture both local and global dependencies within the image for later processing.

The embedded sequence data is then fed into an auto-encoder with a similar structure to that of DDSP [5] where the GRU, as a simplified version of LSTM, helps to extract useful time features from the sequence data. The output of the VDSP Encoder, denoted by f_x, is gone through Fast Fourier Transform (FFT) at discrete time steps n. Let $F[k]$ represent the spectrum of the time-domain feature vector $f_x[n]$ at the frequency $\omega_k = \frac{2\pi}{D}k$. Then, we can extract *magnitude*, denoted by A, and *phase* signals, denoted by ϕ, from F in the frequency domain, respectively. The decomposed magnitude spectrum and phase spectrum vectors are fed into MLP with two fully connected layers (i.e., dense layers) for expanding the representation and capturing more abstract features. The frequency-domain features are then concatenated for subsequent frequency-domain processing.

3.2 Visual DSP (VDSP) Decoder

The VDSP encoder leverages patch embedding combined with FFT to extract more complex frequency-domain features to better understand spatial and contextual relationships in the image data. For instance, analyzing cross-correlation in the spatial domain can be challenging when dealing with two scenery images with the change of a specific semantic property, e.g., more indoor lighting. The spectral correlation analysis reveals the degree to which the two signals share similar frequency components and how they interact at different frequencies and phases, providing a clearer picture of their correlation structure. In this way, we can detect and focus on the prominent frequency features that are most sensitive to the change of the semantic property.

ViT-Based FIR Filter

Core to the VDSP decoder is implementing a linear time invariant (LTI) finite impulse response (FIR) filter that selectively enhances, suppresses, or modifies specific frequency components of the input data with respect to the targeted semantic change. Note that standard convolution layers are equivalent to LTI-FIR filters. Following [6], we employ the frequency sampling method to convert network outputs into impulse responses of linear-phase filters to ensure interpretability and prevent phase distortion.

Instead of manually designing the FIR filter coefficients (e.g., using windowing methods), we propose to train a transformer-like filter module directly from data, such that the output features will contain desired properties learned from referenced images with the targeted semantic attributes. In our context, the response $H(f_x)$ is customized based on explicit control parameters of the filter model trained on the targeted style of the input images.

Specifically, we propose to use a transformer block with self-attention and feed-forward networks to learn the transfer function filter response $H(\cdot)$, as shown in Fig. 1. In particular, the self-attention mechanism in ViT helps learn global frequency dependencies, thus modeling the filter's frequency response across different frequency bands. After the attention mechanism, the frequency representation is passed through a feed-forward (FFN) network, which consists of two ReLu functions with one dense layer in between to introduce nonlinearity. The FFN helps to further process the information captured by the self-attention mechanism, allowing the model to learn more abstract features.

The transformer block processes frequency domain patches with the attention mechanism to learn the global frequency response of the filter by $Z^{(l+1)} =$

Attention$(Q, K, V) + Z^{(l)}$, where $Z^{(l)}$ represents the current frequency domain feature, for $l = 0, 1, 2, \ldots, L-1$, and the attention module captures the inter-frequency dependencies. We empirically use $L = 3$ in our experiments. Additionally, each of these components is followed by Layer Normalization and uses residual connections to stabilize training and improve the flow of gradients.

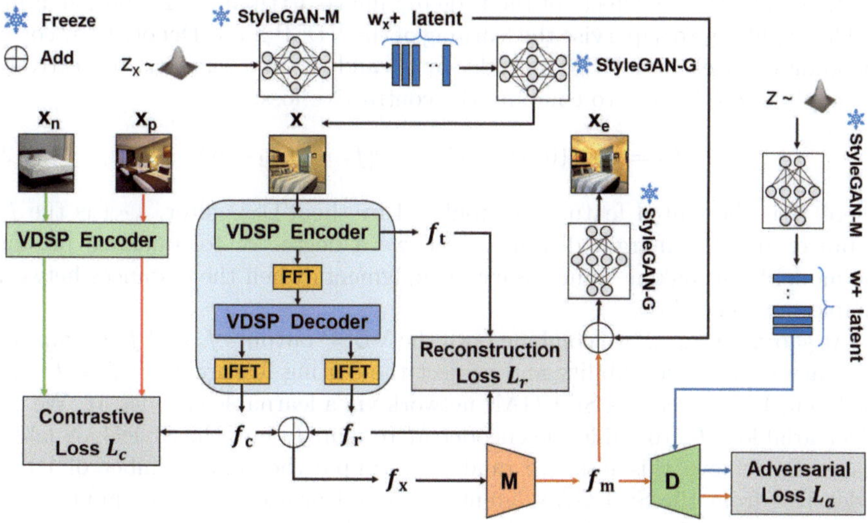

Fig. 2. The overall framework for training the VDSP module with a pre-trained StyleGAN

The filter is trained to impose the desired editing effects on the input frequency components. We then combine the manipulated frequency-domain features (i.e., filtered amplitude spectrum $A'[k]$) with the original phase spectrum $\phi[k]$ and apply the inverse FFT (IFFT) to recover the spatial-domain features. Similarly, we combine $A'[k]$ with the edited phase spectrum $\phi'[k]$ to yield the filter impulse response output f_c. Before applying the inverse FFT (IFFT), the frequency-domain features are passed through two MLP networks, respectively, to help in capturing complex patterns and interactions between features.

3.3 Training Loss Functions

There are three losses for training the proposed network. The first one is the feature *reconstruction* loss, denoted by L_r, comparing the features before and after frequency-domain editing through the VDSP filter. This is to ensure that most information of the image signals is kept unchanged after editing with respect to some specific semantic attribute such as "lighting". Accordingly, we calculate the L_1 distance

$$L_{\text{r}} = \|f_r - f_t\|_1 \tag{1}$$

where f_r is the decoder-recovered feature and f_t is the feature before going through the VDSP filter.

Next is the contrastive loss L_c, which enables a data-driven method for the VDSP filter to learn its transfer function that automatically adjusts the signal components and their correlational effects according to the targeted semantic changes. Specifically, we use a pair of referenced images x_p and x_n that have positive and negative effects of the targeted image attribute, e.g., more lighting and less lighting, to supervise the learning of the VDSP filter. Denote their corresponding features after VDSP encoder by r^+ and r^- for x_p and x_n, respectively. We apply a *Triplet loss* to compute the contrastive loss:

$$L_{\mathrm{c}} = \max\left(0, \|f_c - r^+\|_2 - \|f_c - r^-\|_2 + \alpha\right) \tag{2}$$

where f_c is the edited feature manipulated by the VDSP filter, $\|\cdot\|$ is the L_2 distance, and the margin distance α allows a degree of tolerance to prevent the model from making unnecessary misjudgments when the distances between samples are very close.

As shown in Fig. 2, we combine both the VDSP outputs f_c and f_r to improve the feature expression ability and project the editing feature $f_x = f_c + f_r$ into the latent $W+$ space of a StyleGAN network via a learnable encoder M. We use adversarial loss L_a to guide the encoder M to map f_x into the StyleGAN latent space. To achieve this loss, we randomly sample the same number of latent codes W^+ from the StyleGAN latent space and provide them as input to the discriminator D along with the output f_m of M:

$$L_a^{\mathrm{D}} = -\mathbb{E}_{w\sim Pg(w)}[\log D(W^+)] - \mathbb{E}_{x\sim pX}[\log(1 - D(M(f_x))] \tag{3}$$

$$L_a^{\mathrm{M}} = -\mathbb{E}_{x\sim pX}[\log D(M(f_x))] \tag{4}$$

The adversarial loss is thus $L_a = L_a^{\mathrm{D}} + L_a^{\mathrm{M}}$. Combining the above three loss terms, the overall training object of the proposed method is therefore

$$L_{\mathrm{total}} = \lambda_1 L_{\mathrm{c}} + \lambda_2 L_{\mathrm{r}} + \lambda_3 L_{\mathrm{a}} \tag{5}$$

3.4 Image Editing Process

First, we obtain the modified editing signal f_x through VDSP, and then we map f_x into the latent space of the StyleGAN generator using the encoder M to obtain f_m. Next, we add f_m as the final control signal to the latent code corresponding to the editing target and inject the result into the StyleGAN generator to generate the modified image.

4 Experiment

4.1 Experiment Setting

Dataset: We follow the method in [11] to generate the training set and reference image set using the corresponding pre-trained StyleGAN model. Based on controllable sampling with a fixed random seed, we set the sources of several datasets to be different samplings from the same distribution. Each of the training set and positive/negative reference image sets contains 5000 images. The division of the reference image set is trained using a linear support vector machine (SVM) based on the method in [11] to obtain the linear decision boundary for partitioning the reference set.

Baseline: We compare our method with three StyleGAN-based attribute editing methods: FDIT [9], HyperInvert [12], and FEditNet [13]. All three methods have published their code, and we use the experimental configurations provided by the authors to train the corresponding models, ensuring that the training set used by all methods remains consistent throughout the training process.

Implementation Details: For the VDSP Encoder, we modify the backbone design of the encoder from [6], replacing the module that extracts audio features "MFCC" with the "Patch Embedding" module from Vision Transformer [10]. In this module, we use 16×16 patch embeddings to form input tokens and apply non-overlapping convolutional layers for subsequent downsampling of the tokens. For the VDSP Decoder, we design and implement a LIT-FIR filter with a Vision Transformer (ViT) structure, utilizing multiple layers of standard multi-head attention mechanisms coupled with a feedforward network (FFN) to learn frequency domain features of Magnitude and Phase for modifying frequency components. The pre-trained StyleGAN generator is directly sourced from the officially released pre-trained model [14], and the discriminator follows the design in [15]. All evaluation experiments are conducted using a single NVIDIA Tesla V100 GPU.

Evaluation Metrics: To quantitatively analyze the quality of the editing results and attribute controllability, we use five metrics for editing quality: FID, IS, PSNR, LPIPS and SSIM [16]. We evaluate the image quality on the LSUN-Bedroom and LSUN-Church [17] datasets. For attribute controllability, we follow the setup in [11] and apply rescore analysis to each edited image obtained.

4.2 Performance Evaluation

Figure 3 shows some examples of the image editing results. We use a fixed random seed to ensure that the generated latent codes remain consistent across runs, allowing for a comparison of different methods on the same image. Compared to FDIT [9], HyperInvert [12], our method achieves more accurate attribute control and maintains higher image consistency before and after editing. In contrast to FEditNet[13], our method effectively reflects the interactions among multiple strongly correlated attributes.

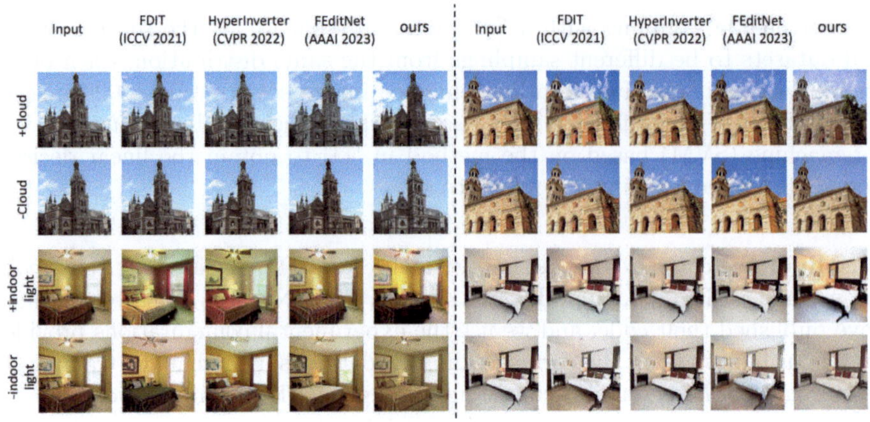

Fig. 3. Comparison of our method's results with those of other image editing methods

Image Quality Comparison: We compare the editing image quality of our method with other methods as shown in Table 1. The results indicate that, across two entirely different scenarios, our method outperforms others in terms of PSNR, LPIPS, and SSIM metrics. This demonstrates that our approach is more effective in maintaining consistency between the edited and original images while achieving the desired editing effects.

Table 1. Image quality comparison with other methods.

Domain	Method	FID (↓)	IS (↑)	PSNR (↑)	LPIPS (↓)	SSIM (↑)
Bedroom	FDIT[9]	40.84	3.25	28.80	0.35	0.93
	FEditNet[13]	43.14	3.26	25.83	0.56	0.89
	HyperInverter[12]	**39.49**	3.18	28.61	0.33	0.92
	Ours	41.11	**3.32**	**29.02**	**0.31**	**0.95**
Church	FDIT[9]	53.52	2.89	27.94	0.47	0.94
	FEditNet[13]	50.84	**3.01**	27.91	0.41	0.91
	HyperInverter[12]	**48.87**	2.79	27.55	0.44	0.92
	Ours	50.43	2.93	**28.83**	**0.39**	**0.95**

Attribute Controllability Analysis: We use re-scoring analysis to demonstrate the controllability of attributes in our method. We apply a pre-trained scene attribute scorer [11] to semantically score the images before and after editing, assessing the changes in attributes to quantify the effects of attribute control. A greater change in a specific attribute score indicates a stronger influence of the editing method on that attribute. Additionally, we showcase the impact of the editing method on non-target but highly correlated attributes, such as cluttered space and Glass semantics in the bedroom scene, and Natural Light and Sunny semantics in the church scene. As shown in Table 2, each column displays the degree of change in semantic scores before and after editing the target attributes using different methods. The greater the change in semantic scores, the stronger the method's influence on that semantic. It is evident that our method achieves the greatest changes in the semantic scores of edited images, indicating a stronger influence on semantics compared to other methods.

Table 2. Attribute controllability analysis by score change.

	Light			Cloud		
	Indoor Light	Glass	Cluttered	Clouds	Natural Light	Sunny
FDIT[9]	0.36	0.15	0.030	0.31	0.25	0.30
FEditNet[13]	0.38	0.17	0.029	0.33	0.26	0.29
HyperInvert[12]	0.25	0.09	0.023	0.20	0.17	0.20
Ours	**0.41**	**0.18**	**0.031**	**0.38**	**0.32**	**0.35**

Ablation Studies: This study employs ablation experiments to validate the significance of each loss term in the model's ability to identify manipulatable semantics. We sequentially remove the contrastive loss L_c, reconstruction loss L_r, and adversarial loss L_a, training the model with the remaining loss terms as optimization objectives before manipulating the indoor lighting semantics. We then utilize a semantic scorer [11] to evaluate the edited samples before and after manipulation, where a greater change in semantic scores indicates a stronger influence of the corresponding loss term on the results. Each configuration utilizes the same hyperparameter settings, and all models are trained for 100 epochs. The results presented in Table 3 demonstrate that the adversarial loss L_a and the contrastive representation loss L_c are particularly crucial. The loss L_a enables the encoder M to map the editing signals extracted by VDSP into the StyleGAN latent space, while L_c determines the model's learning capacity regarding the target semantics.

Table 3. Ablation Study Results

	w/o L_c	w/o L_a	w/o L_r	L_{total}
Indoor Light	0.219	0.092	0.374	0.418
Glass	0.103	0.037	0.167	0.189
Cluttered	0.022	0.004	0.029	0.031

5 Conclusion

In this paper, we propose to develop a more interpretable and controllable approach for image attribute editing by introducing data-driven DSP components to modern image generative models. In particular, we design and implement a visual differential signal processing (VDSP) module for generative image editing within the StyleGAN framework. The paper demonstrates the feasibility of adapting the common DDSP technique (e.g., by incorporating LTI-FIR filters into vision transformers) to improve the generative image processing. Our experimental results show that the proposed method can ensure high fidelity of image editing without compromising the structural integrity of the original input images, and it exhibits strong controllability over specific attributes.

References

1. Ikeda, Y., Doman, K., Mekada, Y., et al.: Lesion image generation using conditional GAN for metastatic liver cancer detection. J. Image Graph. **9**(1), 27–30 (2021)
2. Goodfellow, I., Pouget-Abadie, J., Mirza, M., et al.: Generative adversarial networks. Commun. ACM **63**(11), 139–144 (2020)
3. Shi, Y., Yang, X., Wan, Y., et al.: SemanticStyleGAN: learning compositional generative priors for controllable image synthesis and editing. In: Proceedings of the IEEE/CVF Conference on Computer Vision and Pattern Recognition, pp. 11254–11264 (2022)
4. Karras, T., Laine, S., Aila, T.: A style-based generator architecture for generative adversarial networks. In: Proceedings of the IEEE/CVF Conference on Computer Vision and Pattern Recognition, pp. 4401–4410 (2019)
5. Engel, J., Hantrakul, L., Gu, C., et al.: DDSP: differentiable digital signal processing. arXiv preprint arXiv:2001.04643 (2020)
6. Mirza, M., Osindero, S.: Conditional generative adversarial nets. arXiv preprint arXiv:1411.1784 (2014)
7. Shen, Y., Yang, C., Tang, X., et al.: InterfaceGAN: interpreting the disentangled face representation learned by GANs. IEEE Trans. Pattern Anal. Mach. Intell. **44**(4), 2004–2018 (2020)
8. Pehlivan, H., Dalva, Y., Dundar, A.: Styleres: transforming the residuals for real image editing with StyleGAN. In: Proceedings of the IEEE/CVF Conference on Computer Vision and Pattern Recognition, pp. 1828–1837 (2023)
9. Cai, M., Zhang, H., Huang, H., et al.: Frequency domain image translation: more photo-realistic, better identity-preserving. In: Proceedings of the IEEE/CVF International Conference on Computer Vision, pp. 13930–13940 (2021)
10. Han, K., Wang, Y., Chen, H., et al.: A survey on vision transformer. IEEE Trans. Pattern Anal. Mach. Intell. **45**(1), 87–110 (2022)
11. Yang, C., Shen, Y., Zhou, B.: Semantic hierarchy emerges in deep generative representations for scene synthesis. Int. J. Comput. Vision **129**, 1451–1466 (2021)
12. Dinh, T.M., Tran, A.T., Nguyen, R., et al.: Hyperinverter: improving StyleGAN inversion via hypernetwork. In: Proceedings of the IEEE/CVF conference on computer vision and pattern recognition, pp. 11389–11398 (2022)
13. Xia, M., Shu, Y., Wang, Y., et al.: FeditNet: few-shot editing of latent semantics in GAN spaces. In: Proceedings of the AAAI Conference on Artificial Intelligence, vol. 37, no. 3, pp. 2919–2927 (2023)

14. Karras, T., Laine, S., Aittala, M., et al.: Analyzing and improving the image quality of StyleGAN. In: Proceedings of the IEEE/CVF Conference on Computer Vision and Pattern Recognition, pp. 8110–8119 (2020)
15. Nitzan, Y., Bermano, A., Li, Y., et al.: Face identity disentanglement via latent space mapping. arXiv preprint arXiv:2005.07728 (2020)
16. Lin, J., Li, Z., Tang, X., et al.: VastGaussian: vast 3D gaussians for large scene reconstruction. In: Proceedings of the IEEE/CVF Conference on Computer Vision and Pattern Recognition, pp. 5166–5175 (2024)
17. Yu, F., Seff, A., Zhang, Y., et al.: LSUN: construction of a large-scale image dataset using deep learning with humans in the loop. arXiv preprint arXiv:1506.03365 (2015)

Data-oriented Information System Design and Innovation Management

Design and Implementation of on Orbit Satellite Real Time Detection and Identification System

Keshuai He, Qiang Wu(✉), Jinling Cui, and Xin Zheng

Beijing University of Technology, Beijing, China
{wuqiang,cuijinling,xz}@bjut.edu.cn

Abstract. Facing the real-time processing and high reliability requirements of the intelligent processing load on board, a real-time detection and identification system for on orbit satellites is designed based on the hardware platform composed of Xilinx V7 series FPGA 690T FFG1158-2 and Zhuhai obit Yulong Series CPU Yulong810a. The FPGA in the system receives and forwards the load data, and the Yulong810a in the system runs the neural network to detect, identify and locate the target of the data forwarded by the FPGA. Several state machines are added to the FPGA side to realize real-time state monitoring and multiple handshake protection mechanism, and the process of software annotation of onboard load is optimized. After the test and verification of the ground inspection equipment, the system can meet the high real-time requirements of the on-board system, complete the real-time detection and recognition task and return the target information to the ground.

Keywords: Satellite system · Real time detection and identification · SRIO interconnection system

1 Introduction

In recent years, with the rapid development of commercial space applications, the development requirements of spacecraft payloads are undergoing significant changes. While ensuring high reliability requirements, for the satellite payload, it is more prominent in its processing capacity and adaptability to user needs, low cost, short cycle, batch processing and production.

In 2024, Tianjin 712 Communication and Broadcasting Co., Ltd. proposed the technology of using satellite payload transceiver time compensation method and interval time slot resource planning for the communication delay and poor communication quality of on-orbit satellites, which can ensure the stable and reliable connection of satellite - ground transmission link, and can be perfectly compatible with the technical system design of ground communication equipment. It provides certain feasibility support for the design and implementation of future Leo satellite data link payload [1]. However, this method relies on transceivers to improve the reliability of satellite-ground communication, rather than improving the reliability of payload data transmission by optimizing the system software itself.

A. Ito (Ed.): ICMSSP 2025, LNNS 1637, pp. 139–157, 2025.
https://doi.org/10.1007/978-3-032-05994-9_13

In 2023, Xi'an Branch of China Academy of Space Technology proposed a payload with active multi-beam antennas in view of the higher flexibility requirements of communication satellites gradually changing from high earth orbit satellites to global coverage satellites combining high, medium and low orbits. The payload can be reconfigured in orbit, and the beam size, number and coverage area can be flexibly controlled. Have anti-jamming ability. Generally, mechanical movable antennas or phased array antennas are used in the feeding link to provide services for multiple gate stations [2]. In addition, the research background of the next generation of a certain type of satellite payload design, according to the signal processing requirements and development characteristics of satellite real-time multi-task, the institute designed and implemented a fully digital and universal space-borne signal processing payload architecture. Through the design of hardware, software system, load time-frequency, load power supply, load data flow, load resource management, load reconstruction and load backup, load integration is realized and functional performance requirements are efficiently met [3].

In 2019, the Southwest Institute of Electronic Technology of China designed a reconfigurable on-board signal processing system in view of the characteristics of long development cycle, poor system scalability and weak general ability caused by customized development of satellite electronic payloads. Using modular processing architecture and unified interface specification, and using node dynamic management technology and large bandwidth sampling technology, the hardware reconfiguration of the signal processing system is realized [4]. However, the reconfiguration mode of the system is the traditional dynamic reconfiguration, which relies on the error correction ability of multiple nodes, and no measures are proposed to ensure the stability of communication for a single node.

In 2023, the onboard electronic equipment designed by the 20th Research Institute of China Electronics Science and Technology Corporation uses System-on-Chip (SOC) as the main control chip to realize the software reconstruction function of the system through software version switching, ground update and other software dynamic deployment technologies [5]. In 2022, Xi'an Branch of China Academy of Space Technology proposed a software design architecture to improve the reliability of processor space. The architecture of "Bootloader + main APP + backup APP" was adopted, BootLoader and APP were partitioned, and error correction mechanism was adopted for APP program. Improve the spatial reliability of processors [6].

With the expansion of small satellites in space applications, it is more and more necessary to use CPU to implement complex logic and control functions. However, traditional aerospace grade processors are expensive and bulky, which are not suitable for small satellite systems with small size, light weight and low cost. The processor with small size, low power consumption, low price, high performance and high reliability is an ideal choice for small satellite on-board control equipment. Therefore, it puts forward better requirements for the reliability of file reception [7–12].

This paper mainly focuses on the on-board intelligent processing system based on VPX bus, which needs to have extremely high real-time performance, high-speed serial bus interface and software program registration function. Therefore, the use of traditional FPGA manufacturer recommended official IP core cannot meet the current task of protocol encapsulation analysis and data forwarding function. Therefore, a custom asynchronous UART IP core based on BRAM is proposed in this system, and a highly reliable handshake mechanism is introduced into the system to ensure that the FPGA+ AI chip in the system maximizes the real-time performance of the system under the premise of ensuring the reliability of the system, and relevant tests are carried out for verification.

2 System Software Design

2.1 General Idea of System Software

1) **System requirement analysis**

The software design is mainly based on Xilinx's V7690T as the FPGA processor in the system, which has rich PL end resources. The main process of the system is realized by Xilinx's SRIO IP core, and 6 groups of SRIO are used, including 4 groups on the board and 2 groups off the board. The AXI-Stream bus is used for data processing and forwarding. The program uses Verilog to parse and forward the image data to a specific Yulong810A. The method of handshake between two state machines is used to ensure the correctness of the number of returned results. For the upper part of the program, the traditional AXI UART is replaced by the custom AXI Bram + custom asynchronous UART, the data buffer BUFF is expanded from 16 bytes of AXI UART to 16 KB bytes, and the interrupt trigger mechanism is improved, which greatly reduces the number of interrupt triggers and saves CPU resources. This system uses pure FPGA chip XC7VX690T-2FFG1158I, and the official Microblaze core of Xilinx is added as the system processor in the program. Because of this, the optimization of resources is particularly important.

2) **System requirement analysis**

The implementation block diagram of the hardware platform is shown in Fig. 1. The processor is connected with DDR3 memory and serial port level conversion chip, etc. The system functions involve two parts of FPGA implementation, which are optical load data forwarding based on SRIO and asynchronous UART program registration based on RS422 level. DDR3 is used as a data buffer for program notes. The Microblaze acts as a CPU with an interrupt controller mounted on it for external interrupt triggering. The main logical resources used in FPGA engineering are six groups of SRIO interfaces and one group of MIG.

The SRIO protocol is user-friendly, and the SRIO IP core is instantiated into the project directly, and the data communication can be realized through the AXI-Stream interface. In this system, the NWRITE data type of SRIO is used for data interaction, and the data type uses HELLO package as the separation of data packets, which is convenient for users to parse and encapsulate the protocol.

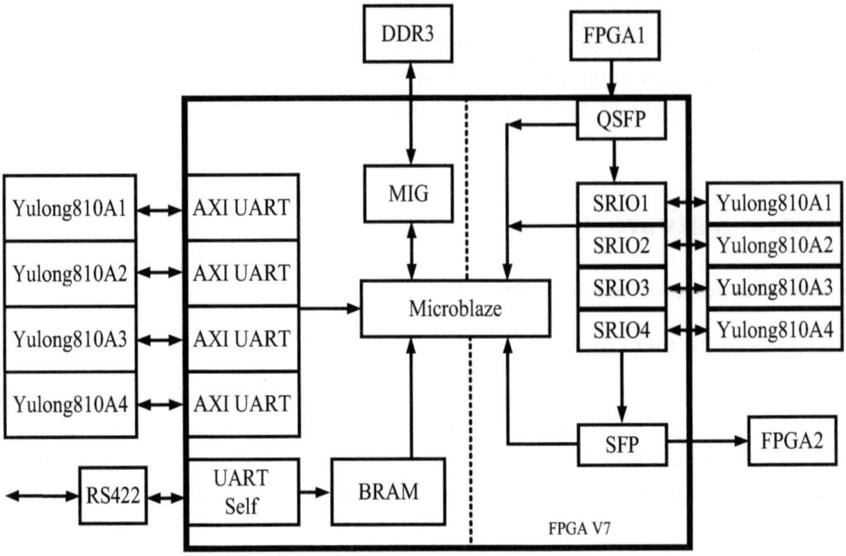

Fig. 1. Block diagram of the engineering platform

2.2 SRIO Interconnection System Design

1) **Design of SRIO clock system**

SRIO initialization module includes four parts: clock, reset and initialization of two GTH modules. FPGA resources are divided into common IO and high-speed transceiver. SRIO is a high-speed signal interface with a speed of 4X 3.125 Gbps. For the Xilinx V7 series FPGA XC7VX690T-2FFG1158I used in this system, there are 32 BUFG clock resources. 7 series devices have 32 global clock lines; These lines can provide clock and control signals to all timing resources of the whole device. Global clock buffers, or BUFGs, are used to drive global clock lines and must be connected to global clock lines.

BUFG directly routes the clock signal to the global clock tree, which can be used in the whole chip. BUFG has strong drive ability, but the clock quality is slightly poor, and the resources are limited. BUFH routes clock signals to its own clock domain and two adjacent clock domains, and its driving ability is second to BUFG, but the clock quality will be better. BUFG can be used as a supplement when it is not enough. Each group of SRIO uses 7 BUFG resources, and the chip cannot meet the clock requirements according to the calculation, so the clock is shared by adjacent banks. The IP core of SRIO is checked in the red box shown in the following figure.

In this system, part of BUFG in SRIO is replaced with BUFH to alleviate the tight clock resources in the current system. After the actual test, the SRIO chain building driven by BUFH is more stable than that driven by BUFG (Fig. 2).

2) **Design of system load data processing**

V7 uses SRIO for external communication, and uses Doorbell+Hello package for external data interaction process. The overall data interaction process is as follows:

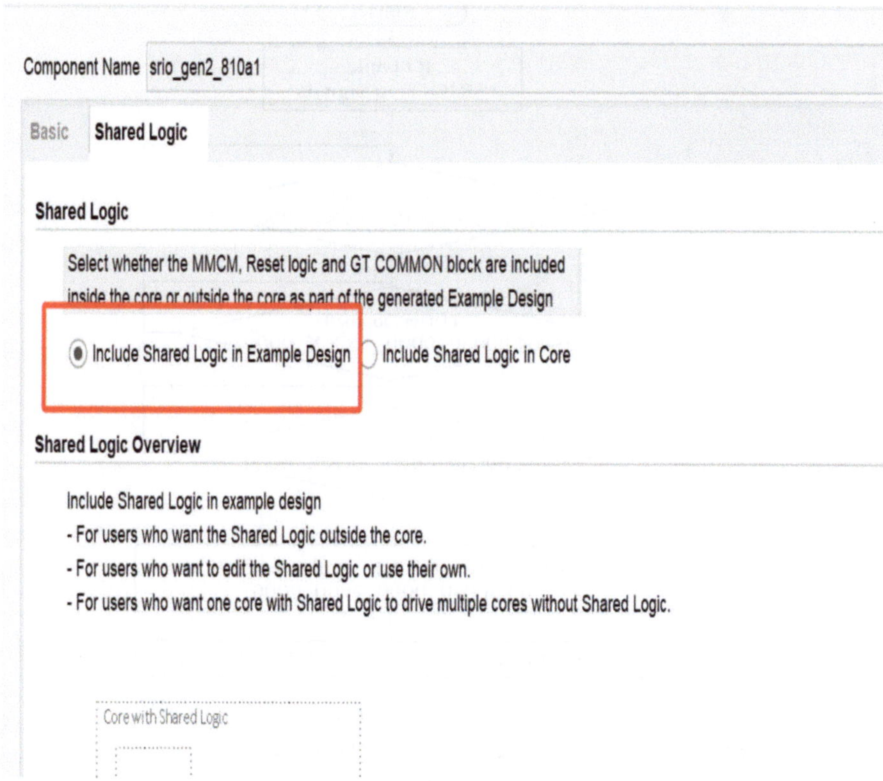

Fig. 2. Screenshot of the SRIO IP core

external sending packet length 1 (doorbell1) - sending packet length 2 (doorbell2) - sending request (doorbell3) - local V7 response is allowed to send (doorbell4) - external sending all slice data of the whole graph. The state machine block diagram is shown below (Fig. 3).

3) **Design of system load data distribution**

In the design of our system, in order to ensure efficient and stable data transmission and processing, specific technical means are used to optimize the whole process.

First, we choose an NWRITE type protocol as the standard for external data sending. NWRITE is a non-blocking write mechanism, which allows the data source to send data without waiting for the acknowledgement of the receiver, which improves the flexibility and efficiency of the data flow to some extent. However, due to its inherent characteristics, the data sending speed under the NWRITE protocol is relatively slow, which is due to the need to ensure data integrity and reliability and avoid data loss when the link is unstable.

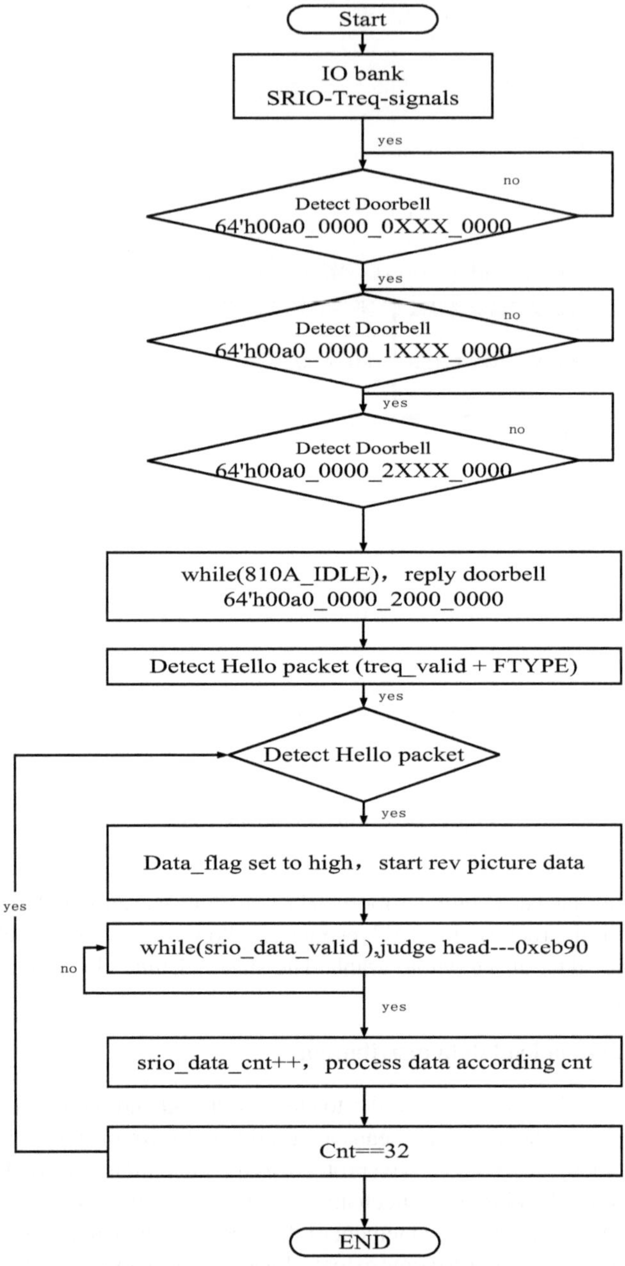

Fig. 3. Flow Chart for Receiving Sliced Images

In order to further improve the system performance, especially in the face of data exchange between different frequency clock domains in the SRIO switching of the current system, we introduce FIFO (First In First Out) buffer. FIFO is a very efficient solution to the problem of data synchronization between asynchronous clock domains. Setting the depth to 16 KB, FIFO can smoothly absorb the impact caused by clock differences and reduce the risk of data overflow, thus ensuring data continuity and integrity. When external data come into the system using the NWRITE protocol, they are held in the FIFO until the target clock domain is ready to receive them, and the ready signal in the AXI-Stream of the target clock domain indicates whether it is in a good receive state.

When the data arrives, the program will dynamically monitor and read the current Serial RapidIO (SRIO) Link status and related working status of each Yulong810A processor. SRIO is a high performance and low latency interconnection technology, which is widely used in embedded systems and communication devices. By monitoring the working status of each Yulong810A chip in real time, the workload of each processor can be accurately understood, and then reasonable task allocation decisions can be made. Specifically, when there is a new image data to be processed, the FPGA will give priority to those idle Yulong810A processors for image detection and recognition. Such a strategy can not only make full use of existing resources to improve the overall processing capacity, but also effectively avoid the increase in latency caused by task stacking.

Determine idle status based on processor status: The system monitors the SRIO Link status and working status of each Yulong810A processor in real-time. When there is new image data that needs to be processed, tasks are prioritized to processors that are completely idle to start task processing as quickly as possible and make full use of resources; At the same time, evaluate the level of busyness of the current work. For processors that are not completely idle but have light loads, based on the progress and expected remaining time of the tasks they are currently processing, as well as the urgency and complexity of the new tasks, comprehensively judge whether the new tasks can be assigned to them. If the processor is expected to complete existing tasks within the time requirements of the new task and be able to process the new task in a timely manner, and the new task will not cause excessive load and performance degradation, then the task can be assigned.

In summary, by using NWRITE protocol, FIFO cache technology and intelligent scheduling mechanism based on SRIO state, the intelligent processing system can achieve efficient image processing task allocation under the premise of ensuring data transmission reliability (Fig. 4).

4) **Design of system load data communication**

Resetting the data cache FIFO after receiving the slice data of an image can ensure that the FIFO will be in a normal state after the completion of a data transmission, which is convenient for the next data transmission. The following state transition diagram shows the state machine process of the FPGA to forward the received data to the Yulong810A (Fig. 5).

Through the Treq interface of SRIO, the data except the doorbell and hello packet headers are stored in the FIFO of the data sending buffer module, and then sent to the external board module (Fig. 6).

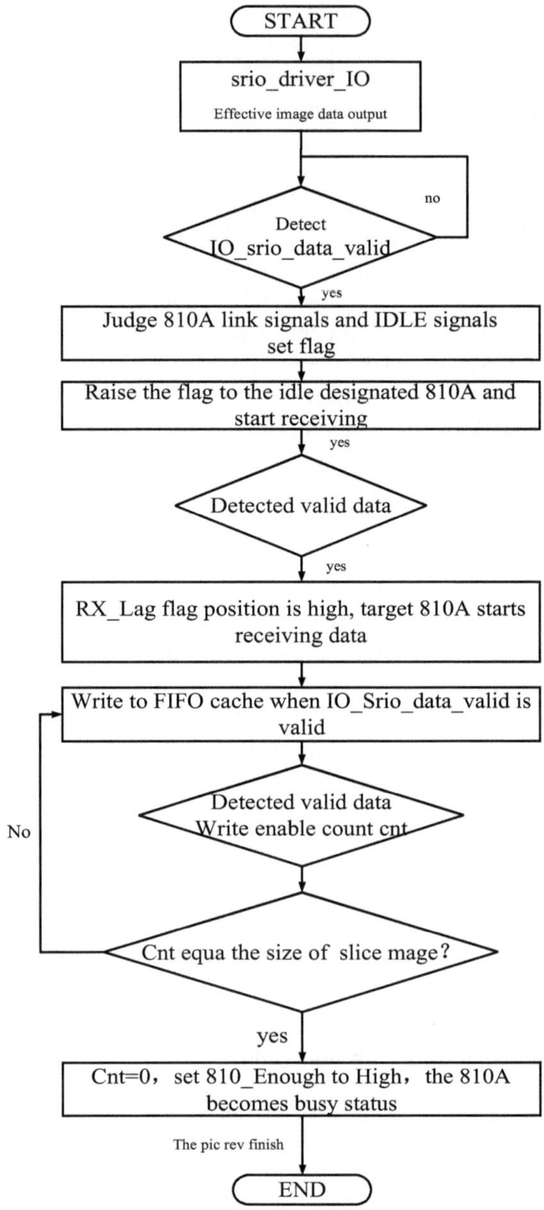

Fig. 4. Chart of Slice Image Distribution

The Yulong810A goes into idle state after receiving the detection and recognition results of an image (Fig. 7).

5) **Design of system detection and identification result return**

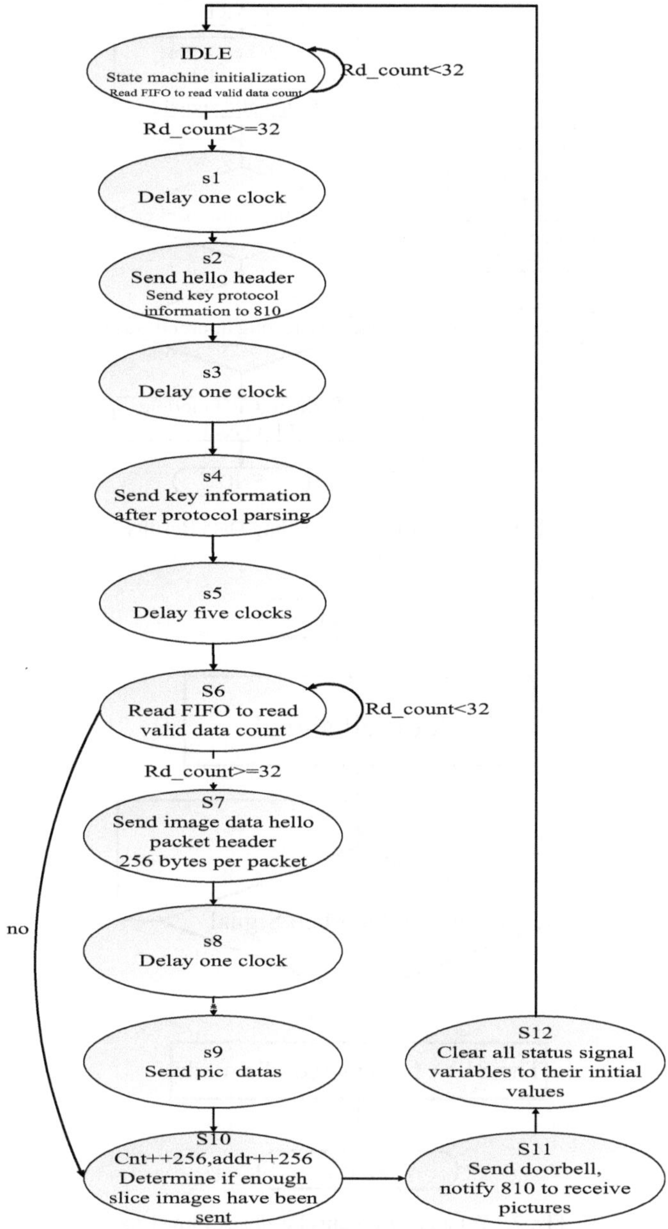

Fig. 5. Data forwarding state transition diagram

After the detection and identification is completed, the result is encapsulated by Yulong810A protocol, and then sent to the external FPGA board through the 2X SRIO. The internal FIFO is used to cache the detection and identification result data.

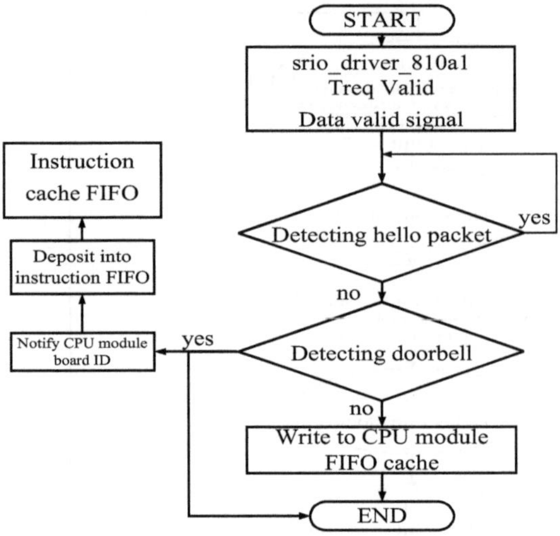

Fig. 6. Flow Chart for Sending Data Results

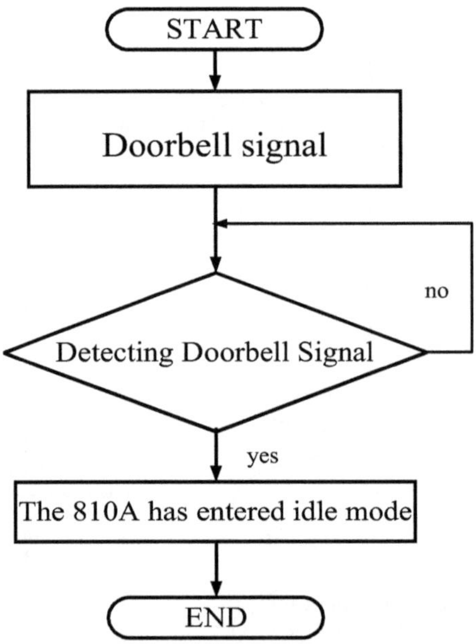

Fig. 7. Yulong810A Status Feedback Flow Chart

The special set instruction FIFO is used to realize the schedule of sending data to the external FPGA, and the data is taken from the FIFO in order to send out the detection and identification results. When any piece of Yulong810A completes the operation, it will send an identification completion signal to the module. When the digital gate circuit captures the rising edge, it reads the sequence number according to the ID of the sent completion signal, and carries out subsequent data processing according to the ID (Fig. 8).

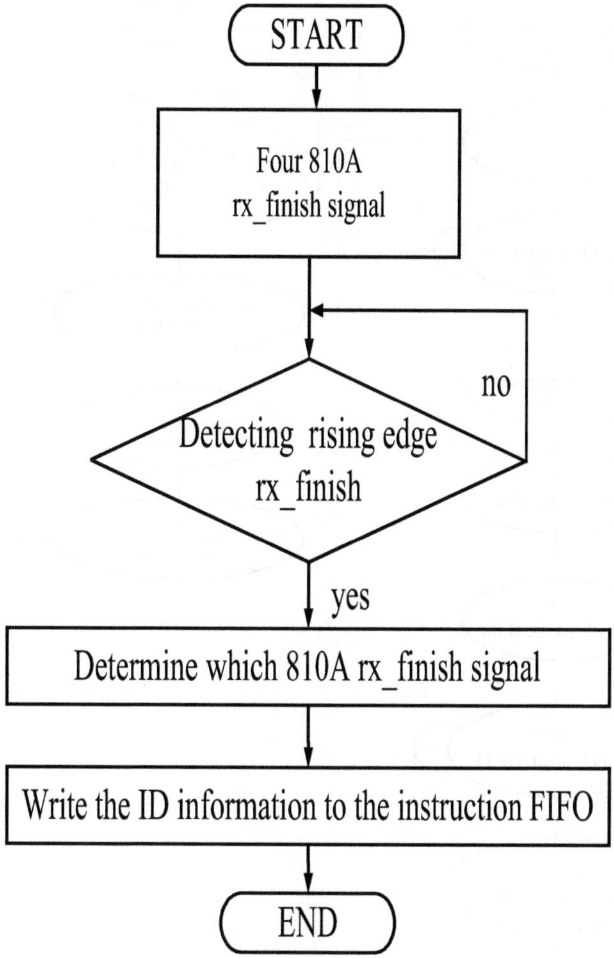

Fig. 8. Yulong810A recognition completion flowchart

The following is the state transition diagram of the two state machines of the module and the relationship between them. State machine 1 is responsible for reading out the ID number in the instruction FIFO, and the relevant flag bit after reading out informs state machine 2 to start the result sending process. State machine 2 is responsible for receiving the state flag bit of state machine 1 and executing the data sending process (Fig. 9).

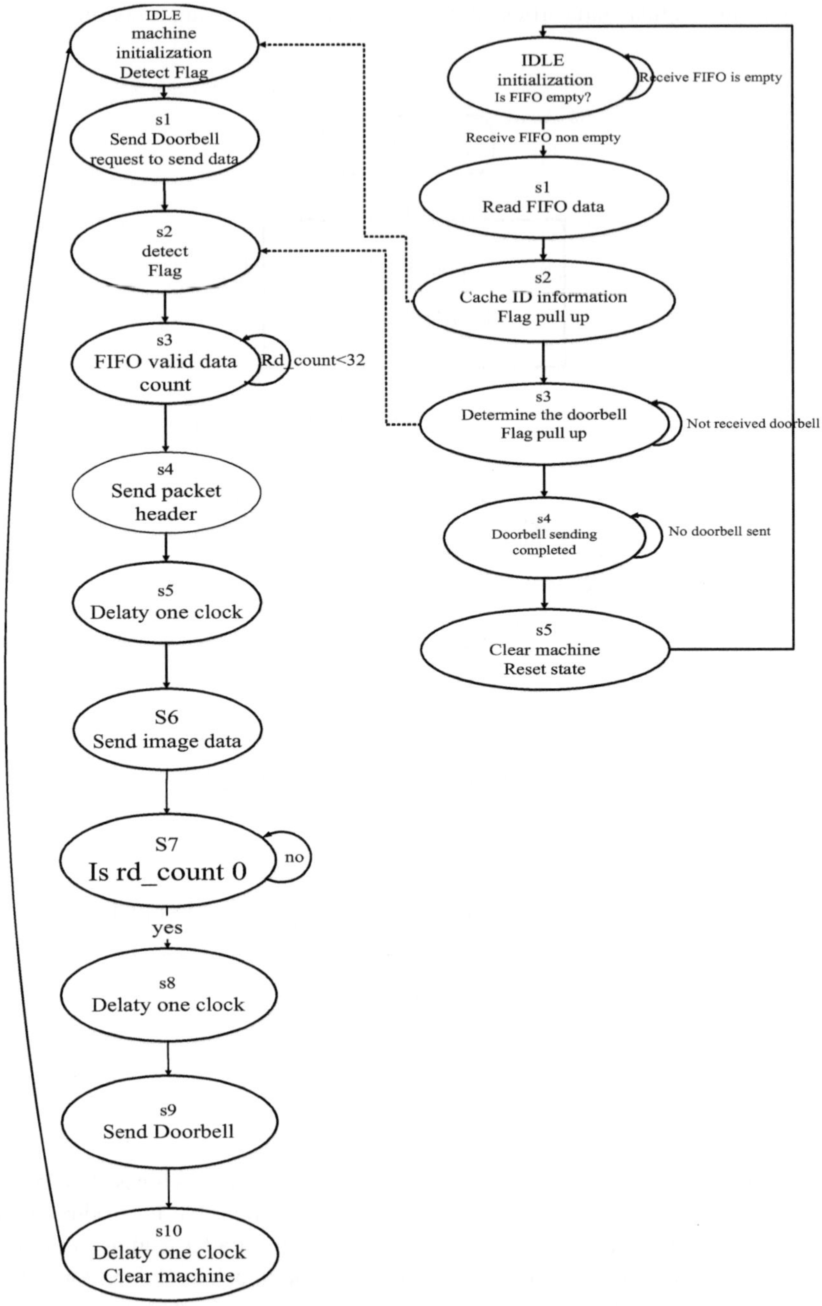

Fig. 9. Yulong810A result feedback state transition diagram

2.3 On Orbit Satellite System Software Annotation Design

1) Define the block diagram of the asynchronous UART structure

For satellites in orbit, the time to build a link is very short. How to efficiently and stably complete the on-orbit update of the program in the limited time has become a very important part. This system proposed a more efficient data transmission mode for the RS422 program registration task, which changed from the commonly used AXI-Lite UART to the custom UART+Bram data transmission mode. Under the premise of ensuring the correct output of the data serial-transfer module, the number of interrupt triggering of a single information frame is reduced. The interrupt of the system is implemented by Microblzae + AXI Intr interrupt controller, and the main frequency clock of Microblaze is 100 MHz. The optimized architecture can greatly reduce the interrupt handling pressure of Microblaze.

AXI Lite UART is triggered by a single byte interrupt. When the data processing pressure is high, the system software may lose data or fail to respond to other interrupts. In this system, FIFO is added as the data buffer after the serial-transfer and parallel module, the depth is adjusted to 8 KB and the ping-pong buffer is added.

When the cumulative time exceeds the preset time and no data arrives, an interrupt is initiated to Microblaze. The soft core completes the data transfer in the interrupt service function. The measured moving time of 8KB data with cache open is milliseconds, which can meet the performance and real-time requirements of the scene (Fig. 10).

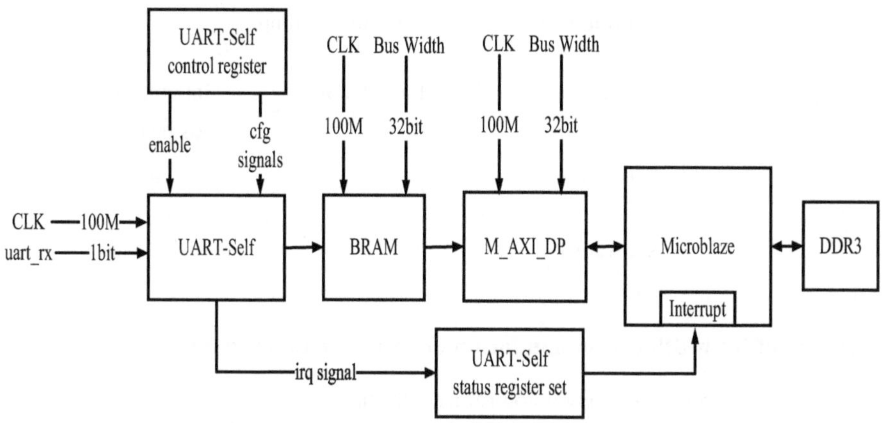

Fig. 10. UART module block diagram

The custom UART module completes the status register configuration through the AXI Lite bus, including module baud rate, data start bit, stop bit, idle interrupt time interval, etc. After the configuration module is enabled, the data is moved, and the software interrupt in the program does not need to be cleaned up.

2) **Interrupt design of on orbit detection system**

The current data interrupt response mode is the idle interrupt trigger, that is, after the completion of data serialization and concatenation, there is still no data written to the FIFO after a period of time, then the interrupt is triggered. The backend uses 100M clock to perform FIFO state polling, and writes BRAM operation when the empty signal of FIFO is not high. Since the data parsing clock domain of the system serial and parallel module is 100M, but the data transmission and parsing rate is relatively low compared with the main frequency, the main state of the backend state machine is idle state most of the time. We only need to ensure that the FIFO cache of the data domain is large enough to accommodate the write data rate of the back-end BRAM. Each time the serial and parallel module of the custom UART finished parsing a data, it would refresh the accumulated value of the timer. When the value met the value set by the user, the state machine would jump to the state of interrupt trigger, and notify Microbale to take the data from the ping-pong buffer. Ping-pong caching also ensures that there are no memory exceptions where read and write Pointers point to the same memory area. There are five external interrupts in the program, which are software interrupts of FPGA, four Yulong810A chips in the system, and software interrupts on the program (Table 1).

Table 1. Microblaze Software Interrupts.

Interrupt Num	Interrupt Name	Interrupt SGI	Interrupt Type
0	Software 0	0	Rising edge
1	Software 1	1	Rising edge
......	......	......	......
4	Software 4	4	Rising edge

3) **Design of bit width conversion for on orbit maintenance data**

Add the data bit width conversion register in the program state machine, the single byte data after string conversion and module analysis is arranged in the order of the size end, the data uses the wire type variable to complete the data assignment and writes to the BRAM interface in accordance with the shift register, which solves the problem that the single byte conversion of four bytes of bit width conversion needs to be completed by FIFO. The system resources are saved. The bit-width conversion flow chart is shown in the following figure (Fig. 11).

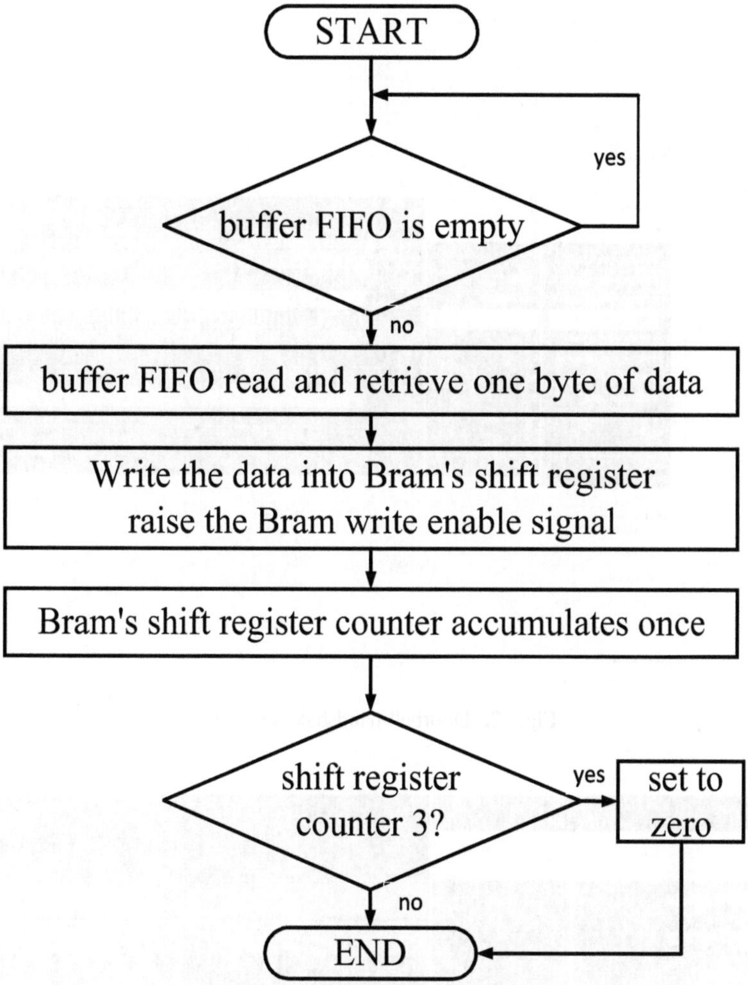

Fig. 11. Flow Chart for Bit Width Conversion

3 Test Result

This system was compiled and debugged using Vivado 2018.3 and Xilinx SDK 2018.3. Tests were conducted on the system's image reception, image forwarding, and result feedback scheduling. The following are the test results (Figs. 12, 13, 14, 15, 16 and 17).

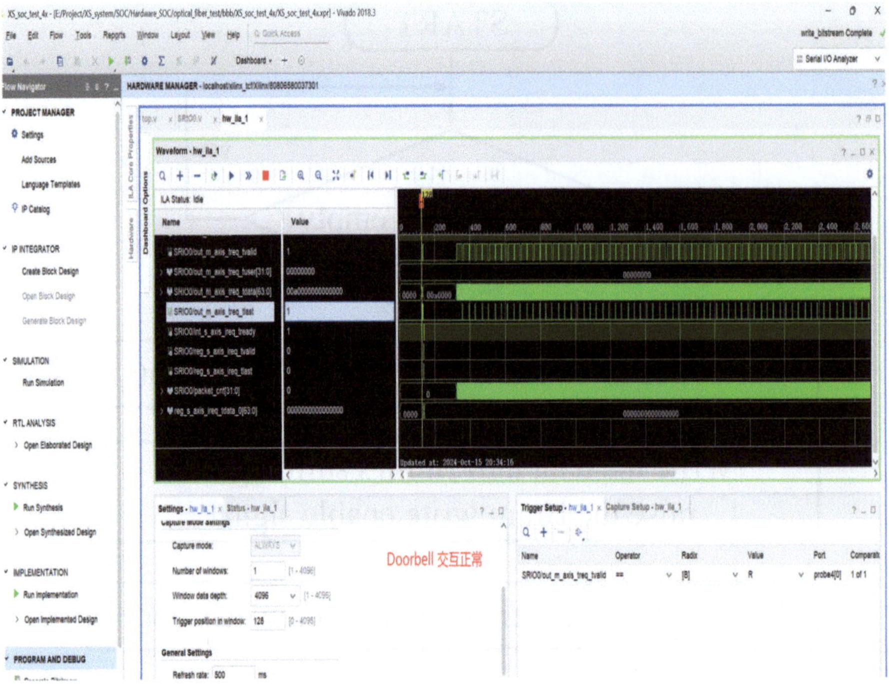

Fig. 12. Doorbell handshake diagram

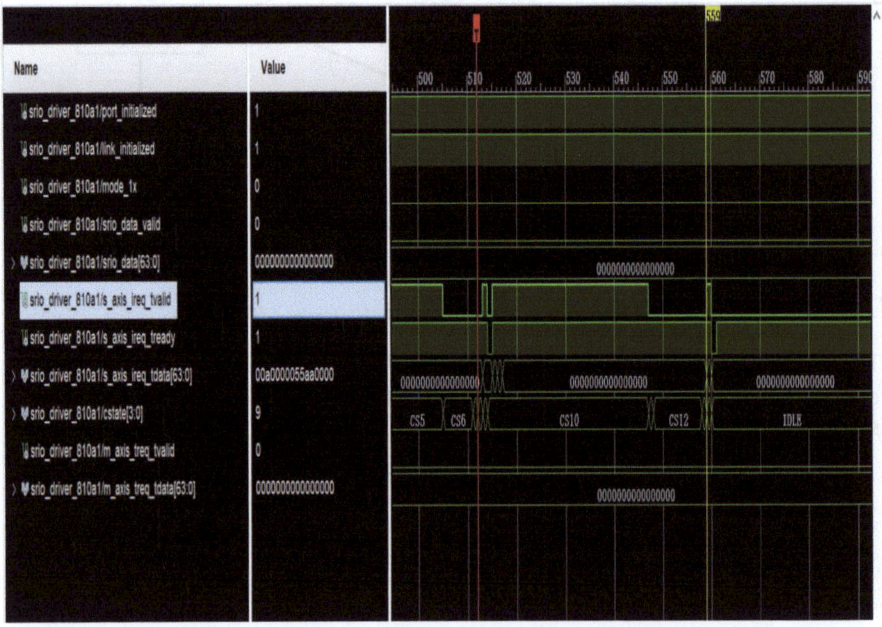

Fig. 13. Communication diagram between FPGA and Yulong810A

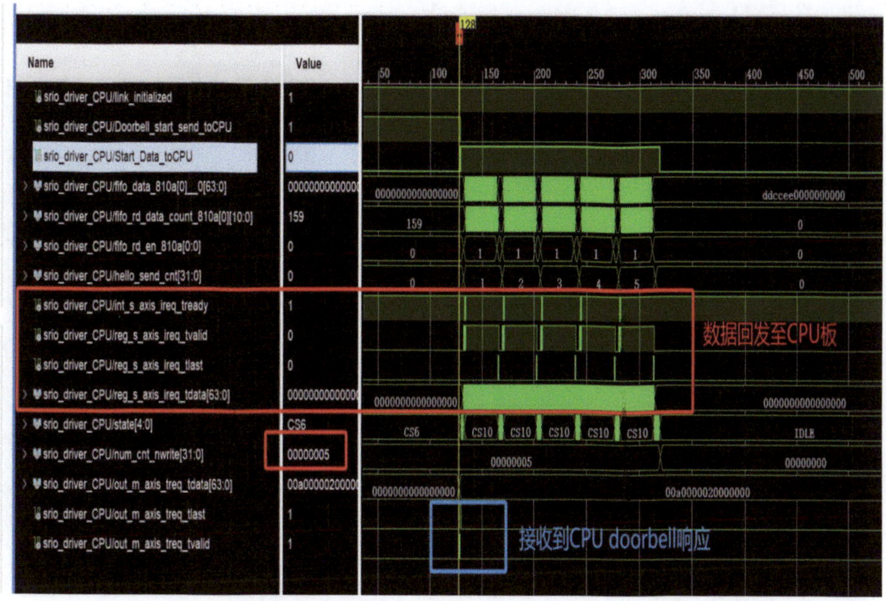

Fig. 14. Communication diagram between FPGA and external FPGA

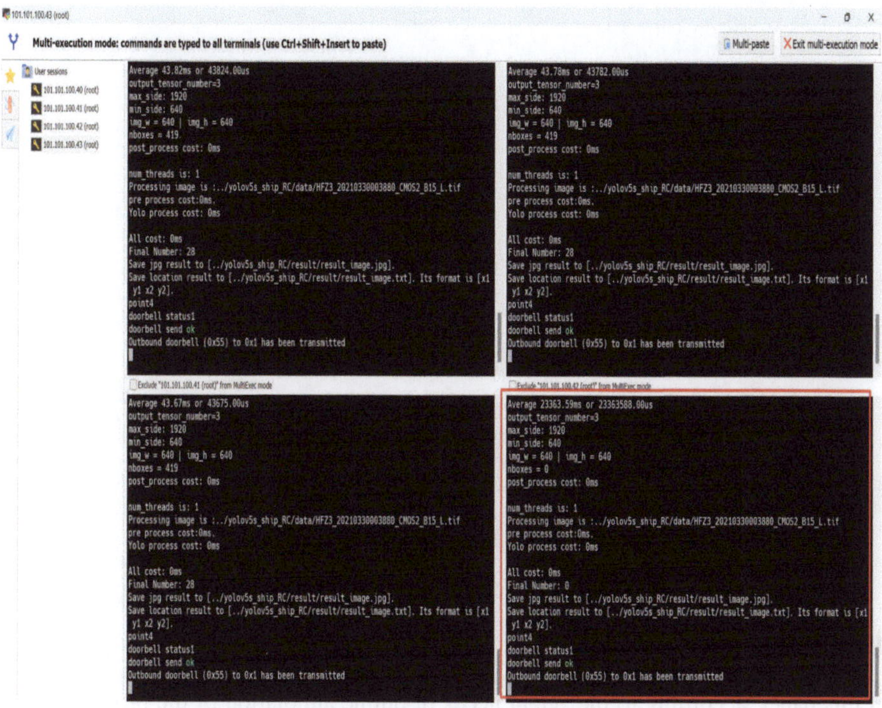

Fig. 15. Yulong810A detection and identification diagram

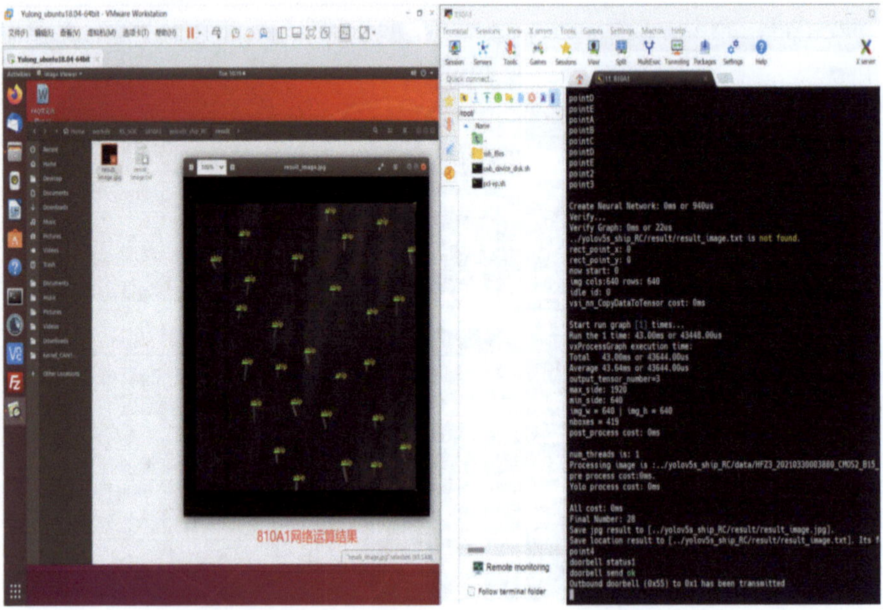

Fig. 16. Yulong810A Detection operation results diagram

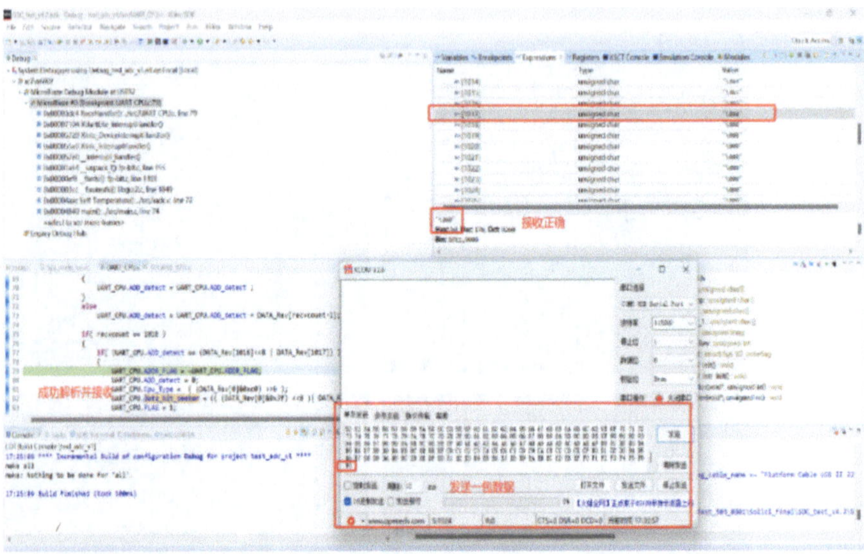

Fig. 17. Doorbell handshake diagram

4 Conclusion

In this paper, according to the actual needs of online annotation of the program of the on orbit load processing system, a heterogeneous platform of FPGA+CPU is designed based on the hardware platform composed of Xilinx V7 series fpaga 690T, FFG1158-2 and Zhuhai obit Yulong Series CPU Yulong810A, which realizes the real-time detection and identification of satellite loads on the premise of ensuring the real-time performance. Aiming at the problem that MicroBlaze soft core is difficult to complete complex system functions as the processor in the system, the optimization design is carried out, and the system control and status monitoring are realized. The process optimization is carried out for the annotation part of the onboard load program software, which greatly reduces the number of MicroBlaze interrupts, so as to ensure that the system can meet the high real-time requirements of the on-star system.

References

1. Wang, G., Song, W.: Research on delay measurement and compensation technology of low orbit satellite data link. In: Tianjin Electronics Industry Association. Proceedings of 2024 Annual meeting of Tianjin Electronics Industry Association. Tianjin 712 Communication Broadcasting Co., Ltd., June 2024. https://doi.org/10.26914/Arthurc.nkihy.2024.009905
2. Zhou, Y., Kang, D., Lou, D., et al.: Review on flexible load technology of communication satellite. Space Electron. Technol. **20**(03), 29–38 (2023). (in Chinese)
3. Wang, L., Wang, Y.: Design of load architecture for full digital spaceborne signal processing. Space Electron. Technol. **20**(02), 76–83 (2023). (in Chinese)
4. Ma, C.: Design of reconfigurable signal processing system for satellite electronic loads. Electron. Des. Eng. **14**(23), 35–39 (2019)
5. Zhang, D.: Realization of software in-orbit reconstruction function based on SOC. Mod. Navig. **14**(03), 205–210 (2023)
6. Du, X., Zhong, R., Li, Q., et al.: Design and implementation of ground simulation system for intelligent processing of remote sensing image on satellite. Remote Sens. Technol. Appl. **37**(05), 1198–1208 (2022)
7. James, A., Agrawal, A.K.: Spatially resolved chemiluminescence imaging of multiple combustion species using a single high-speed camera. Meas. Sci. Technol. (2023)
8. Holladay, S., Zhang, Z.: Linear and 2D arrays for focused laser differential interferometry using a high-speed camera. Opt. Commun. **546**, 129754 (2023)
9. Xing, Z., Zhu, S., Huang, Xi., et al.: Software defined spacecraft system architecture design. Spacecr. Eng. **30**(5), 1–8 (2019). (in Chinese)
10. Fan, Z., Dong, S., Lu, F., et al.: Design of a space-borne multi-function reconfigurable terminal. Comput. Meas. Control **30**(4), 149–154 (2002)
11. Wang, S., Zhang, S., Huang, X., et al.: Monolithic multi-processing architecture for space-borne SAR imaging and Intelligent processing. J. Northwest. Polytech. Univ. **39**(3), 510–520 (2021)
12. Song, T., Du, C., Qiu, X., et al.: Design and implementation of spatial reliability software architecture based on ARM processor. Space Electron. Technol. **19**(1), 44–48 (2019)

Information Systems Planning for IT Service Management of Higher Education Institution

Prisilla Amiel A. Sarto(✉), Helaena Marie B. Bobis, Patrick Leus M. Vidal, and Grace Lorraine D. Intal

School of Information Technology, Mapúa University, Manila, Philippines
{paasarto,hmbbobis,plmvidal}@mymail.mapua.edu.ph, gldintal@mapua.edu.ph

Abstract. Effective management of computer laboratories is crucial for providing seamless services to students, faculty, and staff in a Higher Education Institution (HEI). The researchers investigated the synergistic application of ITIL 4, Process Mining techniques, and Artificial Intelligence (AI) tools within the IT Service Management (ITSM) of an HEI in the Philippines. The focus lied specifically on Incident Management and Service Request Fulfillment on computer labs. By using the data gained from conducting interviews to the HEI's ITSM staff, analyses such as process mapping, SWOT/TOWS matrix, and gap analysis were performed. Through the combined results of the analysis tools, the researchers were able to determine two areas in the current ITM process that needs improvement: Response Time and Data Management. Following the declared research objectives, an implementation plan was proposed by the researchers. Value Stream Processes for Incident Management and Service Request Management served as the blueprint for said implementation plan, both of which highlights inclusion for Service Desk Operations. The blueprints both highlighted the implementation of ITIL 4 practices and Process Mining with AI, both of which were supported by a review of related literatures. An Action Plan going over the details of the actual feasible implementation such as the project's description, scope, and scheduling was also proposed. The researchers then made a theoretical conclusion that with an emphasis to response time reduction and user satisfaction, the service desk should serve as the central hub for IT support, and that utilization of SLAs, OLAs, and UCs ensured a standardized approach to service delivery.

Keywords: Higher Education Institution (HEI) · ITIL 4 · Process Mining techniques · Artificial Intelligence (AI) · IT Service Management (ITSM)

1 Introduction

ITIL is a globally recognized framework for IT Service Management (ITSM) that allows organizations to align IT services with business strategies to focus on creating value for end-users. Its benefits and structured approach are why organizations adopt this framework [2]. Together, the ITIL 4 incident management, service request management, and service desk practices ensure that the organization "quickly restores service when

A. Ito (Ed.): ICMSSP 2025, LNNS 1637, pp. 158–173, 2025.
https://doi.org/10.1007/978-3-032-05994-9_14

it fails, effectively handles service-related requests, and provides the communication channels for the above and all other service-related communications" [1].

Adapting this framework is essential for the HEI as IT serves as leverage for advancement in aligning to the mission, vision, and objectives of the HEI to become globally competitive which makes a university a business organization that operates in the education field [10]. The service desk acts as a medium between service providers and users with incidents and service requests as the main concern [1]. With that, it is important to note that these three practices go hand in hand with each other, and if ITSM doesn't have these services in full swing, it can significantly impact business performance and financial outcomes [5].

In this aspect, there are automation technologies that can improve ITSM such as Process Mining [25] and Artificial Intelligence (AI). Over the last decades, Process Mining has been a novel technology that helps businesses to better understand their processes through the benefits that it brings such as organized and accurate data management and the automated process insights from event log data and identification of different factors and variants of a business process [4]. On the other hand, AI has proven its importance in businesses by revolutionizing decision-making processes by providing accurate and valuable insights with massive amounts of data in an extremely short time [22]. This paper presents an IS/IT strategic plan aimed at improving the efficiency of the HEI's ITSM in incident management, service request management, and service desk operations through Process Mining and AI. The researchers have deemed this strategy efficient as it breaks down complex processes, such as the ITSM of an HEI, into manageable, step-by-step procedures for targeted improvements [24].

2 Business Analysis

The researchers have interviewed the IT department head of an HEI to analyze how they handle incidents and service requests. To align their plans with their outcomes, the researchers also surveyed students who use the computer laboratories about their experiences regarding the response of the IT department to their situation with incidents and service requests.

2.1 Service Blueprint

The researchers have selected a service blueprint as a process mapping tool to enhance clarity and streamline communication within the IT department's operations. Service blueprints map both user interactions and internal processes, highlighting key touchpoints and areas in need of improvement [3]. They effectively identify inefficiencies and support prioritization of operational improvements, as seen in other fields like healthcare and education [14, 15]. Given their adaptability, service blueprints provide a practical framework for optimizing department-specific workflows and addressing unique process challenges [23, 26].

Leveraging the clarity and structure provided by the service blueprint, the researchers identified key weaknesses in the IT department's current processes. Through interviews

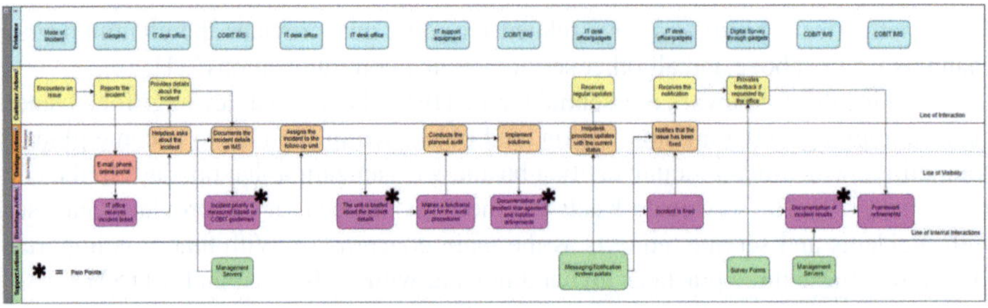

Fig. 1. Service Blueprint of the HEI's Incident and Service Request Management in Computer Labs.

and surveys, they gathered detailed information, enabling Fig. 1 to outline the department's pain points. The blueprint revealed challenges such as the IT team's reliance on manual file handling for lab reservations and attendance sheets, complicating data storage and traceability. It also highlighted students' lack of awareness regarding service standards, making it difficult to report issues without disrupting classes, and noted that prolonged incident resolution times contribute to significant classroom interruptions.

With that, the service blueprint provided essential insights for the IS/IT strategic plan, clarifying current operations and identifying areas for automation and improvement. This mapping directly guided targeted enhancements within the department's processes.

3 IS/IT Current Situation

3.1 Understanding the Current Situation

Process Framework. The HEI's ITSM does not utilize the best and most recognized framework for IT services. Their incident management and service request management lack a rigid framework like ITIL. Their process doesn't possess such structure which leads to hindrances in value creation to its users.

Service Desk. Since they are not able to implement ITIL, they do not practice service desk. Their help desk still needs improvement, and this hinders communication between the users and service providers regarding incidents and service requests.

Business Technology Stack. The HEI's ITSM also does not utilize existing technological advantages that are available for them in order to increase their efficiency. Their printed forms should transition to e-forms for users to fill out in issuing requests and logging in incidents and to seamlessly record it into their database. This will provide them an easier and more accurate data management which will be helpful for Process Mining and AI integration in the future.

3.2 Assessment of Current Strategies

SWOT Analysis. The HEI's strengths give it a competitive advantage for opportunities. As seen in Table 1, the first strength of the HEI lies in its experienced staff. With HEI's strict employment, they were able to hire competent professionals who are open to growth and can adapt to such advancements. This also made it possible for them to handle all the incidents and service requests for all the computer laboratories in the HEI. This is tough considering that these computer laboratories are used by hundreds of students every day while they still deliver a wide range of services within their service catalog. On the other hand, the company's weaknesses give it a disadvantage. Their response time isn't quick enough causing a disruption in classes. Human errors because of printed loose files lead to inefficiencies and data errors.

The HEI can take advantage of a range of opportunities for growth. Their potential involves implementing process automation. The HEI can take advantage of a range of opportunities for growth. Their potential involves implementing process automation tools like Process Mining and AI which can also help them with data management. They can also delve into having self-service options for their defined small issues so users can have a more personalized experience and lessen response time. On the other hand, threats impede the HEI's progress. As they implement new services and the HEI grows, naturally, more incidents and service requests will need to be fulfilled leading to an increase in workloads. Because of the massive amounts of data produced, data can be difficult to handle and may lead to security breaches (Table 2).

Table 1. SWOT Analysis of the HEI's Incident and Service Request Management in Computer Labs

STRENGTHS	WEAKNESSES
Experienced staff members Handles all incidents and service requests for all the computer labs. Wide range of services in the catalog	Response time for resolution and request fulfillment causes class disruptions. Printed loose files lead to inefficiencies and data errors.
OPPORTUNITIES	**THREATS**
Implementing process automation tools Promoting self-service options for users in simple issues Leveraging AI for data management	Increasing workload as incidents and requests may rapidly surge as the HEI grows. Security breaches pose a threat to data integrity

TOWS Matrix. Through this strategic management tool, the researchers were able to combine elements of a SWOT to generate actionable strategies. By using the TOWS matrix, organizations like HEIs, can leverage internal strengths to overcome threats posed by various risk categories observed internationally [8]. The HEI's strengths position is to capitalize on opportunities. With a highly experienced staff and a robust service desk, the HEI can leverage its expertise to innovate and improve incident handling and service delivery. This skilled workforce enables the HEI to manage a comprehensive service catalog efficiently, ensuring users have access to a diverse range of offerings, thereby enhancing overall satisfaction.

Table 2. TOWS Matrix/Strategies for the HEI's Incident and Service Request Management in Computer Labs

STRENGTHS/OPPORTUNITIES STRATEGIES	**WEAKNESSES/OPPORTUNITIES STRATEGIES**
S1, O1, O3) Leverage experienced staff and a robust service desk to explore new opportunities. S1, S2, O2) Use the broad service catalog to promote additional services and expand existing ones.	W1, O1, O2) Improve response time by investing in automation tools and self-service options. W2, O3) AI integration to enhance incident detection and data management
STRENGTHS/THREATS STRATEGIES	**WEAKNESSES/THREATS STRATEGIES**
S1, T2) Mitigate security risks by utilizing experienced staff and implementing robust security measures. S2, S3, T1) Address increasing workload by optimizing existing processes and automating where possible.	W2, T1, T2) Mitigate data errors by addressing rapid surge of workload and security risks. W1, T2) Ensure regulatory compliance by leveraging experienced staff and implementing necessary controls.

However, the HEI also faces weaknesses that could hinder its ability to seize opportunities. Notably, its response time is inadequate, causing disruptions in classes and overall service efficiency. Additionally, reliance on manual processes contributes to human errors and data inaccuracies, posing further challenges. To address these weaknesses and capitalize on opportunities, the HEI can invest in automation tools to improve response times and streamline incident resolution and request fulfillment processes. Integration of artificial intelligence (AI) for incident detection and data management can enhance accuracy and efficiency, paving the way for better service outcomes.

Furthermore, the HEI must also confront external threats that could impede its progress. As the institution grows and introduces new services, the workload will inevitably increase, potentially straining existing resources. Moreover, the proliferation of data presents challenges in data management and security, with the risk of breaches looming large. To mitigate these threats, the HEI can optimize existing processes to manage workloads effectively and implement stringent security measures, leveraging its experienced staff to enforce necessary controls and ensure regulatory compliance. By addressing weaknesses, seizing opportunities, and mitigating threats, the HEI can navigate its incident and service request management landscape more effectively, ensuring continued growth and success.

Gap Analysis. Table 3 shows the researchers' PESTLE analysis after conducting the SWOT/TOWS analysis for the HEI. This is to identify the HEI's gap in different sectors of society. This is important because it prepares the organization for market shifts, identifies risks, and aligns strategies with the changing environment [19]. The HEI should consider the Political factors that influence incident management and service requests, with opportunities arising from favorable government policies and threats from potential regulatory changes. Economic factors like growth provide resource allocation opportunities but budget constraints during downturns pose threats. Social factors highlight opportunities in meeting user expectations but challenges in addressing changing demographics. Technology offers opportunities through digital transformation but threats from cyber security risks. Environmental factors present sustainability opportunities but threats from natural disasters. Legal factors offer compliance opportunities but threats from legal challenges like lawsuits.

Table 3. PESTLE Analysis for the HEI's Incident and Service Request Management in Computer Labs

POLITICAL	**ECONOMIC**
1) Explore opportunities from favorable policies related to education, technology adoption, and research funding. 2) Monitor potential regulation changes like data privacy laws.	1) Leverage economic growth to allocate resources effectively for incident management. 2) Economic downturns may affect funding for IT services.
SOCIAL	**TECHNOLOGICAL**
1) Understand user preferences and tailor incident handling to meet their needs. 2) Address the diverse needs of students, faculty, and staff.	1) Embrace technology advancements for efficient incident tracking and resolution. 2) Protect against data breaches and cyber threats.
ENVIRONMENTAL	**LEGAL**
1) Align incident management practices with environmental sustainability goals. 2) Prepare for disruptions caused by environmental events.	1) Ensure adherence to legal requirements like accessibility standards. 2) Mitigate risks related to lawsuits or non-compliance.

The SWOT, TOWS, and PESTLE analyses serve as the current state assessment of the HEI's incident management and service request processes. SWOT pinpoints internal strengths (e.g., experienced staff) and weaknesses (e.g., manual processes), alongside external opportunities (e.g., process automation) and threats (e.g., increasing workload). TOWS then integrates these factors, generating actionable strategies such as leveraging staff expertise for innovation and addressing workload challenges through automation. PESTLE delves into macro-environmental influences, aiding in anticipating changes and aligning strategies with industry trends [19].

By combining these analyses, the researchers were able to identify areas for improvement:

- Response Time: The current average response time does not meet the expectations of students and faculty, leading to disruptions in classes. Research shows that organizations with optimized ITSM processes can achieve faster resolution times, enhancing user satisfaction [12].
- Data Management: The HEI faces challenges in managing large volumes of data, which can lead to security breaches. Implementing robust data quality assurance

methods is essential for mitigating these risks. Studies indicate that effective data governance frameworks can significantly reduce the likelihood of data breaches [6].

These identified gaps have helped the researchers identify the desired state of the HEI's ITSM which is through Automation and AI Integration: The objective is to implement process automation tools like Process Mining and AI to enhance data management and reduce response times. Recent studies indicate that automation can significantly improve IT Service Management (ITSM) by streamlining processes and minimizing human error [16]. Specifics of this are explained in the next section.

Research Objectives. Through the thorough analysis of the researchers of the current situation of the HEI, the primary objectives of this research are as follows:

- *Investigate the Integration of ITIL 4, Process Mining, and AI:* This objective aims to explore how these methodologies can synergistically enhance IT Service Management (ITSM) within a higher education institution (HEI) in the Philippines, specifically focusing on incident handling and service request fulfillment in computer labs.
- *Develop an IS/IT Strategic Plan:* The research seeks to create a comprehensive strategic plan for incident management, service request management, and service desk processes that aligns with the institution's mission and vision.
- *Assess Current Processes and Identify Gaps:* Through interviews and analyses such as SWOT/TOWS matrix and gap analysis, the research intends to assess the current ITSM processes, identify weaknesses, and recommend improvements.
- *Promote Continuous Improvement:* The objective includes establishing recommendations for data quality assurance methods and continuous process improvement to enhance user experiences.

The researchers have developed a strategic implementation plan aligned with these objectives, which will be detailed in the subsequent sections of the paper.

4 IS/IT Direction and Strategies

4.1 Review of Related Literature

ITIL in HEIs' Service Delivery. The University of Oxford's IT Services department, under the leadership of Andrew Dixon, successfully implemented ITIL 4 to enhance its service delivery. By adopting ITIL 4, they achieved a holistic view of their IT operations, enabling them to move to a more mature and professional IT management approach. The implementation of ITIL 4 allowed the teams to understand the interdependencies of their services, make continuous improvements, and align their efforts toward a common goal [20]. This study provides researchers insights into how ITIL 4 is well-suited for HEIs' ITSM due to its ability to provide fit-for-purpose solutions that co-create value consistently. ITIL 4 helped the University of Oxford's IT Services department to shift to a more agile and responsive mode of operation, especially crucial during the Covid-19 pandemic [20]. This proves that ITIL 4 helps achieve HEIs' objectives of adaptability and aligning themselves globally.

ITIL in HEIs' Service Desk. The University of Oxford embarked on an IT Transformation program to enhance service levels for its IT users by consolidating three independent help desks into a single service desk. This is case study is extremely helpful for the researchers as they also aim to implement ITIL service desk practices into the HEI. Pink Elephant, a global service management specialist, conducted assessments, identified best practices, and defined requirements for the new service desk. They also assisted in selecting an ITSM toolset aligned with ITIL principles, ensuring a smooth transition and effective implementation. The utilization of ITIL-aligned processes and frameworks by Pink Elephant showcased the benefits of ITIL in HEIs' ITSM service desk. By adopting ITIL practices, the University of Oxford was able to streamline its IT services, improve efficiency, and provide a reliable service desk. The structured approach of ITIL helped in defining processes, selecting appropriate tools, and ensuring that the desired way of working was reflected in the implemented solution [21]. This is significant for this study as it provides insights into how service desk practices in ITIL's incident and service request management are optimal for this study.

Process Mining on Service Desk. This study by Tello et al. focuses on utilizing process mining tools, which is one of the goals of the researchers. The study demonstrates how analyzing event logs can lead to insights for improving processes already in place. The study showcases the benefits of process mining for the service desk by providing a structured approach to incident resolution, rapid restoration of business processes, and continuous monitoring for agile service recovery [25]. This study is significant for this research as it shows how process mining can extract valuable knowledge from event data, enabling organizations to discover, monitor, and enhance processes effectively, ultimately improving the Incident Management process at the service desk.

Combination of Process Mining and AI. Microsoft's Power Automate Process Mining integrates process mining capabilities with next-generation AI. This offers researchers a powerful tool for optimizing incident and service request management which are all aimed by this study. By extracting event data and utilizing AI for root cause analysis, it identifies process inefficiencies and automatable tasks, enabling resource optimization and efficient data management. Customizable reports aid in KPI monitoring and compliance, while their cost-effective implementation within the Microsoft ecosystem streamlines adoption [11]. Researchers find this very helpful as they can integrate both Process Mining and AI into one in the IS/IT strategic plan. The researchers see the benefits of this regarding ease of implementation and cost-benefit.

4.2 Project Content and Specifications

Through the thorough RRL of the researchers, the aim for the HEI is to adopt ITIL 4, Process Mining, and AI in their process. The project aims to boost productivity and efficiency in their incident and service request management. To further this, the HEI must establish a robust service desk and implement Process Mining with AI in its data management. This project aligns with the HEI's objectives of streamlining IT services to further their growth and global competitiveness. The researchers recommend the HEI invest in such as this project aligns with the HEI's objectives of streamlining IT services to further their growth and global competitiveness.

4.3 Data Quality Assurance and User Experience Recommendations

Data quality assurance is essential for enhancing user experience and supporting IT Service Management (ITSM) goals. By implementing specific approaches such as automated data validation techniques, robust data governance practices, and interface enhancements, organizations can significantly improve data accuracy and user interaction. The following sections outline these strategies in detail.

Automated Data Validation Techniques

- Real-time Validation: Implement automated checks during data entry to ensure accuracy and completeness.
- Data Profiling: Regularly assess data quality metrics to identify anomalies and inconsistencies [13].
- Error Detection Algorithms: Utilizing machine learning models to flag potential data errors before they impact decision-making [7].

Data Governance Practices

- Establishing Policies: Creating clear data governance frameworks that define roles, responsibilities, and standards for data management [17].
- Continuous Monitoring: Implementing feedback loops and root cause analysis to maintain high data quality over time [13].
- Data Stewardship: Assigning data stewards to oversee data integrity and compliance with governance policies [7].

Interface Enhancements for User Interaction

- User-Centric Design: Developing intuitive interfaces that facilitate easy data access and manipulation, enhancing user satisfaction [18].
- Feedback Mechanisms: Incorporating user feedback tools to continuously improve the interface based on user experiences [18].
- Training and Support: Provide resources and training on data quality tools and practices to help users understand and engage effectively with the system

By implementing these strategies, the HEI's ITSM system can achieve higher data quality and improved user experience, which are critical to effective incident and service request management.

5 Implementation Plan

5.1 Value Stream Process for Incident Management

Figure 2 shows the proposed value stream process of the researchers to the HEI that involves the implementation of ITIL 4 for incident management, with the Service Desk serving as the central point for reporting, sorting, resolving, and logging incidents. Following the resolution of each incident, the process incorporates Process Mining with

AI, enabling the extraction of valuable data insights from incident logs. This simplified value stream process of incident management is extracted from Axelos and enhanced by the researchers through Process Mining with AI [1]. In the lower part of the diagram, the different ITIL 4 management practices besides incident and service request management are observed. The service level management (SLM) offers direction on service levels for incident management, information security management guarantees that only authorized users report incidents, and knowledge management provides the techniques and knowledge necessary for diagnosis and resolution [1]. These insights are stored in the HEI's database, supporting enhanced operational efficiency, data-driven decision-making, and a continuous improvement cycle. This integration aligns incident management operations with long-term strategic objectives, showcases an institutional commitment to innovation, and provides a competitive edge in delivering IT services.

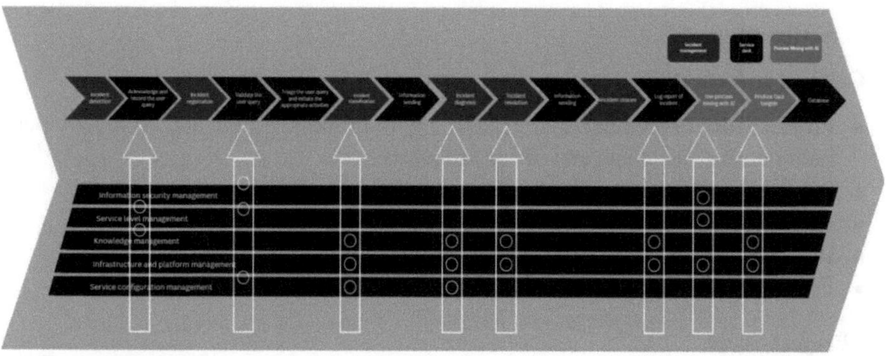

Fig. 2. Proposed Value Stream Process with Process Mining and AI for the HEI's Incident Management and Service Desk.

5.2 Value Stream Process for Service Request Management

Figure 3 shows the proposed value stream process of the researchers to the HEI that involves the implementation of ITIL 4 for service request management, with the Service Desk serving as the central point for accepting requests, validating queries, fulfilling service requests, gathering feedback, and logging fulfilled requests. Following the fulfillment of each service request the process incorporates Process Mining with AI, enabling the extraction of valuable data insights from service fulfillment logs. This proposed service request serves a similar purpose and almost has the same process as the proposed incident management process. It also utilizes the different ITIL 4 management practices, as observed in the lower part of the diagram. These practices collaborate synergistically toward a shared objective of establishing a resilient and enhanced system for the HEI.

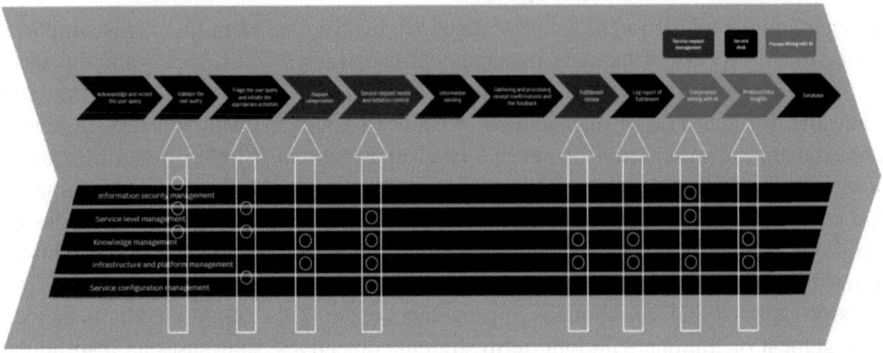

Fig. 3. Proposed Value Stream Process with Process Mining and AI for HEI's Service Request Management and Service Desk.

5.3 Action Plan

I. **General Information**

- Project Title: Implementation of ITIL 4 with Process Mining and AI in Computer Laboratories' incidents and service requests
- Project Duration: 3 months
- Beneficiaries: IT department (Computer Labs)
- Estimated Project Cost: 350,000 Pesos [power price]
- Project Lead: IT Department Head

II. **Project Scope**

1. Project Goals and Objectives:

   - Adopt ITIL 4 practices to enhance incident management, service request handling, and overall IT service delivery.
   - Streamline processes, reduce response times, and improve user satisfaction by implementing ITIL 4 with process mining techniques enhanced by AI.
   - Analyze logs and use produced data insights to enhance efficiency and decision-making.
   - Create a centralized service desk as a single communication point for incident and service request handling.
   - Leverage IT to stay globally competitive and ensure IT services support the HEI's growth.

2. Project Requirements:

   - The implementation must establish a service desk that can accommodate users' incident reports and service requests.
   - The implementation must successfully log reports and produce insights from using Process Mining with AI.
   - The implementation must provide a faster target time for incident resolution and service request fulfillment through the service desk.
   - The implementation must observe efficient data management through the service desk.

3. Project Description

The IT department managing computer labs can handle all the incident reports and service requests issued by the students, faculty members, and staff through the service desk. The IT department staff must solve the incidents and fulfill requests according to the Service Level Agreements (SLAs) and Operational Level Agreements (OLAs) that are established once the ITIL 4 practice is already implemented. The targeted time stated in SLAs and OLAs must be faster than the previous resolution time before ITIL 4 implementation. All the steps during the process of resolving incidents and fulfilling service requests must be logged and communicated through the service desk. The data acquired from logs must be processed by the chosen software of the HEI that has process mining with AI. The produced insights must be stored in the database of the service desk to aid in enhancing efficiency and decision-making.

4. Project Deliverables

- Requirement Specification
- ITIL 4 Certification from Axelos
- Software provider for Process Mining with AI
- Publishing of SLM SLAs, OLAs, and Underpinning Contracts (UCs) for the IT department's service delivery
- Finalized Value Stream Process for Incident Management and Service Request Management with utilization of Service Desk
- Established Service Desk
- Integration of Service Desk data to the IT department's database
- AI training using service desk data to produce data insights
- Notice to users of the enhanced process of incident reporting and service request query

6 Project Schedule

The course of the project will be research, development of the process, finalization of service requirements, training of AI, and execution. The IT department head and team must first research how they will implement ITIL 4 in their current process based on the researcher's recommendations. They must decide the scope and limitations of the implementation in the computer labs. After finalizing, they must obtain certification from Axelos on the use of ITIL 4 in their organization. The IT team must then develop a process through the proposed value stream process of the researchers which they will use in the implementation. Then, they must research software providers that offer Process Mining with AI. As per the RRL, the researchers recommend that the HEI look into Microsoft's Power Automate Process Mining. This stage must be accomplished in not more than 30 days.

Next, the HEI must finalize its service requirements by publishing its service level management SLAs, OLAs, and UCs, for a well-defined IT service delivery. This will make the process standardized and easy to follow for the IT staff. This stage must be accomplished shortly after the first stage. The researchers recommend that this stage be finished in not more than three weeks or 21 days.

Then, they must test run the software acquired by training the AI with the existing dataset of resolved incidents and fulfilled service request logs from the IT department's database. This is to ensure that the AI will produce valuable insights for the IT department to use. This should be finished rather quickly in not more than 14 days.

Lastly, the execution wherein the IT departments will inform students, faculty members, and other staff about the implementation of the process. The whole execution process should be achieved in not more than three weeks or 21 days. This already includes the adjustment period of IT service providers and users within the HEI. In total, the researchers recommend that the project be accomplished in the scheduled 3 months.

6.1 Alignment with Research Objectives

The researchers have aligned the proposed strategies with the research objectives by:

- *Integrating ITIL 4, Process Mining, and AI:* This strategy directly supports the objective to enhance IT Service Management (ITSM) by leveraging established frameworks and technologies to improve incident handling and service request fulfillment.
- *Developing a Comprehensive IS/IT Strategic Plan*: The strategic plan is designed to address the objective of creating a structured approach for incident and service request management, ensuring alignment with the institution's mission and vision.
- *Conducting Thorough Assessments of Current Processes:* By employing tools such as SWOT/TOWS analysis and gap analysis, the researchers can identify weaknesses in current practices, thereby aligning their strategies with the objective of assessing and improving existing processes.
- *Implementing Continuous Improvement Mechanisms:* Recommendations for data quality assurance and process improvements are aligned with the objective of enhancing user experiences, ensuring that strategies focus on ongoing development and responsiveness to user needs.
- *Utilizing Automation Tools:* The proposed strategies include investing in automation technologies to streamline processes, which aligns with the objective of improving efficiency in incident management and service request fulfillment.
- *Enhancing Communication through a Robust Service Desk:* This strategy supports the objective of fostering better communication between service providers and users, which is essential for effective incident management.

By aligning these strategies with the research objectives, the study aims to create a cohesive framework that enhances ITSM within the higher education institution, ultimately leading to improved service delivery and user satisfaction.

6.2 Data Quality Assurance and User Experience Recommendations

Each following recommendation supports the HEI's goals for consistent, high-quality data and a user-friendly experience:

Data Quality Improvement Methods.

- Data Governance Framework: Establish policies and standards for data management to ensure consistency and accountability [13].
- Data Profiling and Auditing: Regularly assess data quality using profiling techniques to identify inaccuracies and inconsistencies, ensuring high data quality across all processes [13, 27].
- Data Cleansing and Enrichment: Implement tools to cleanse data by removing duplicates and enriching datasets for completeness, optimizing data relevance and accuracy [9, 13].
- Continuous Monitoring: Use metrics and dashboards to monitor data quality over time, enabling timely interventions when needed [13, 28].

User Training and Engagement.

- Training Programs: Regularly train staff on best practices in data management, emphasizing the importance of data quality for system effectiveness [27].
- Feedback Loops: Establish channels for users to report data quality issues, encouraging a culture of continuous improvement and responsiveness [13].
- Service Desk Optimization: Enhance the service desk interface to facilitate easy reporting of data issues and to streamline user interactions with IT support [9].

While these strategies aim to improve data quality, recognizing that user experience can be affected by the complexity of data management systems is needed. Simplifying interfaces and providing user-friendly tools are crucial to enhancing engagement and satisfaction.

7 Conclusion

By understanding the current situation of HEI's Incident Management, Service Request Management, and their corresponding Service Desks, the researchers can now propose a theoretical implementation plan that demonstrates a strategic approach to improving the higher education institution's IT service management (ITSM) by integrating ITIL 4, Process Mining, and AI. By establishing structured processes for incident and service request management, the HEI aims to reduce response times and improve user satisfaction, positioning the Service Desk as a central hub for IT support. The integration of process mining with AI offers data-driven insights that enhance decision-making and efficiency, aligning operational practices with long-term institutional goals. Moreover, the use of Service Level Agreements (SLAs), Operational Level Agreements (OLAs), and Underpinning Contracts (UCs) ensures a standardized approach to service delivery, fostering greater transparency and accountability. Additionally, the plan emphasizes continuous improvement and data quality through various measures, including data governance, cleansing, and monitoring. User engagement is also prioritized, with training programs and optimized service desk interfaces to facilitate seamless interactions and feedback loops for IT staff and end-users alike. Together, these initiatives contribute to a robust and resilient ITSM framework that not only should improve daily operations but also reinforce the HEI's commitment to innovation, user satisfaction, and strategic alignment with its broader educational objectives.

References

1. Adyrbai, D.: ITIL Practices in 2000 words: incident management, service desk and service request management, 1 July 2021. https://www.axelos.com/resource-hub/white-paper/itil-2000-words-incident-service-desk-request-management
2. Axelos: What is ITIL®? | Axelos (2019). https://www.axelos.com/certifications/itil-service-management/what-is-itil/
3. Bartch, V.M., Wolf, T.L.V., Lee, S.A., Poncelet, S.A., Nemec, S.L., Morgenthaler, T.I.: A service blueprint approach to prioritize operational improvements in a new outpatient clinic. Healthcare **11**(4), 100715 (2023). https://doi.org/10.1016/j.hjdsi.2023.100715
4. Brocke, J.V., Jans, M., Mendling, J., Reijers, H.A.: A five-level framework for research on process mining. Bus. Inf. Syst. Eng. **63**(5), 483–490 (2021). https://doi.org/10.1007/s12599-021-00718-8
5. Bughin, J., Van Zeebroeck, N.: The Best Response to Digital Disruption. McKinsey & Company (2017). https://www.mckinsey.com/mgi/overview/in-the-news/the-right-response-to-digital-disruption
6. Chukwurah, N., Ige, A.B., Adebayo, V.I., Eyieyien, O.G.: Frameworks for effective data governance: best practices, challenges, and implementation strategies across industries. Comput. Sci. IT Res. J. **5**(7), 1666–1679 (2024). https://doi.org/10.51594/csitrj.v5i7.1351
7. Clarke, N.: How to ensure provision of accurate data to enhance decision-making. J. Secur. Oper. (2019). https://doi.org/10.69554/tkla5594
8. Dandage, R.V., Mantha, S.S., Rane, S.B.: Strategy development using TOWS matrix for international project risk management based on prioritization of risk categories. Int. J. Manag. Proj. Bus. **12**(4), 1003–1029 (2019). https://doi.org/10.1108/ijmpb-07-2018-0128
9. Devineni, S.K.: Data quality improvement strategies for data analysis specific to the IT industry. J. Artif. Intell. Cloud Comput., 1–7 (2023). https://doi.org/10.47363/jaicc/2023(2)189
10. Fuada, S.: Incident management of information technology in the Indonesia higher education based on COBIT framework: a review. EAI Endorsed Trans. Energy Web, 156387 (2018). https://doi.org/10.4108/eai.13-7-2018.156387
11. Graham, J.: Microsoft launches Power Automate Process Mining and next-generation AI. Microsoft Power Platform Blog, 18 July 2023. https://www.microsoft.com/en-us/power-platform/blog/2023/07/18/microsoft-launches-power-automate-process-mining-and-next-generation-ai/
12. Hidayatullah, D., Munir, M., Sobandi, A., Furqon, C.: The effect of business process reengineering (BPR) and service quality with technology utilization on educational institutional performance through stakeholder satisfaction. ATESTASI Jurnal Ilmiah Akuntansi **7**(2), 1026–1042 (2024). https://doi.org/10.57178/atestasi.v7i2.905
13. Idemudia, C., Ige, A.B., Adebayo, V.I., Eyieyien, O.G.: Enhancing data quality through comprehensive governance: methodologies, tools, and continuous improvement techniques. Comput. Sci. IT Res. J. **5**(7), 1680–1694 (2024). https://doi.org/10.51594/csitrj.v5i7.1352
14. Kim, D., Cho, J.: Analysis of the health examination service process using service blueprint: focus on the older adult patient in South Korea. Healthcare **11**(20), 2709 (2023). https://doi.org/10.3390/healthcare11202709
15. Kim, Y.S., He, J., Seppälä, L.: A service blueprint approach in ship building activity mapping. In: IMDC 2024 (2024). https://doi.org/10.59490/imdc.2024.827
16. Kumar, R., Kawin Singh, V., Harish, K., Bhavish, R.K.: Importance of intelligent automation in post COVID era: a study. In: 2021 2nd International Conference on Computation, Automation and Knowledge Management (ICCAKM), pp. 204–209. IEEE (2021). https://doi.org/10.1109/ICCAKM50778.2021.9357745

17. Kumar, B.S., et al.: Data integrity and data governance. In: IntechOpen eBooks (2022). https://doi.org/10.5772/intechopen.100778
18. Macfarlane, S.B., AbouZahr, C.: A matter of trust: data quality and information integrity. In: Palgrave Macmillan UK eBooks, pp. 427–449 (2019). https://doi.org/10.1057/978-1-137-54984-6_22
19. Makos, J.: Advantages of PESTLE Analysis in Business Strategic Planning. PESTLE Analysis, 21 May 2024. https://pestleanalysis.com/advantages-of-pestle-analysis/
20. Oxford University: The University of Oxford and ITIL case study. Axelos, 7 August 2020. https://www.axelos.com/resource-hub/case-study/university-of-oxford-and-itil-case-study
21. Pink Elephant Development: Service Desk transformation at the University of Oxford - Pink Elephant United Kingdom. Pink Elephant United Kingdom, 31 October 2024. https://pinkelephant.co.uk/about/case-studies/service-management-case-study/
22. Prasanth, A., Vadakkan, D.J., Surendran, P., Thomas, B.: Role of artificial intelligence and business decision making. Int. J. Adv. Comput. Sci. Appl. **14**(6) (2023). https://doi.org/10.14569/ijacsa.2023.01406103
23. Silva, H.M., et al.: Service blueprint of comprehensive medication management: a mapping for outpatient clinics. Res. Soc. Adm. Pharm. **17**(10), 1727–1736 (2021). https://doi.org/10.1016/j.sapharm.2021.01.006
24. Sirkemaa, S.J.: Information systems management – understanding modular approach. J. Adv. Inf. Technol. **10**(4), 148–151 (2019). https://doi.org/10.12720/jait.10.4.148-151
25. Tello, I., Ruiz, C., Yoo, S.G.: Analysis of COBIT 5 process "DSS02 - manage service requests and incidents" for the service desk using process mining. In: 2018 International Conference on eDemocracy & eGovernment (ICEDEG), pp. 304–310 (2018). https://doi.org/10.1109/icedeg.2018.8372335
26. Wernke, D., Teixeira, J.M.: Visual Management: mapping experiences from service blueprints in the remote scenario. Modapalavra E-periódico **14**(31) (2021). https://doi.org/10.5965/1982615x14312021244
27. Yang, K.: Data Quality, pp. 199–236. Wiley (2023). https://doi.org/10.1002/9781119932475.ch6
28. Zuo, X.: Research on data quality improvement program based on big data application. IEEE Xplore (2023). https://doi.org/10.1109/iciba56860.2023.10165495

VEXine: Automating SBOM and VEX Generation Using Transformer LLM Models with LangGraph

Su-Yeon Yoon[1], Min-Ju Kwon[1], Sang-Beom Kim[1], Youn-Gee Kim[1], Jong-Won Park[1], Seung-Ho Son[2](✉), Young-Ok Kim[3](✉), and Si-Hyeong Jeong[1](✉)

[1] KITRI (Korea Information Technology Research Institute) BoB (Best of the Best), Seoul, Republic of Korea
shjeong.in.korea@gmail.com
[2] SKKU (Sungkyunkwan University), Seoul, Republic of Korea
seungho96.official@gmail.com
[3] HD Hyundai, Seongnam, Republic of Korea
brizcreg@gmail.com

Abstract. As software systems grow more complex and supply chain attacks become more advanced, ensuring security and transparency is a global priority. This paper introduces VEXine, an AI-powered framework that automates the generation of Software Bill of Materials (SBOM) and Vulnerability Exploitability Exchange (VEX) reports using transformer-based large language models. By combining advanced AI with automated security assessment tools, VEXine enhances the accuracy, efficiency, and scalability of supply chain vulnerability management. Evaluations against traditional software composition analysis (SCA) tools show that VEXine delivers superior performance, providing actionable insights and addressing critical security challenges in modern software ecosystems.

Keywords: SBOM · VEX · LLM · LangGraph

1 Introduction

The increasing complexity of software systems and the rise of supply chain attacks make it essential to ensure software security and transparency. The Log4j vulnerability [1] and the SolarWinds Orion attack highlight the risks in the supply chain, significantly affecting over 400 Fortune 500 companies and government agencies. Supply chain attacks exploit trust relationships, threatening public trust, individual privacy, and national security. These incidents demonstrate the need for a more proactive and transparent approach to software vulnerability management. Particularly, during the COVID-19 pandemic, supply chain attacks increased by 650%, driven by rapid digital transformation and growing reliance on the software ecosystem. Unlike traditional attacks, supply chain attacks amplify their impact by exploiting trust between software developers, distributors, and end users. The SolarWinds Orion attack compromised over 400 Fortune 500

A. Ito (Ed.): ICMSSP 2025, LNNS 1637, pp. 174–186, 2025.
https://doi.org/10.1007/978-3-032-05994-9_15

companies and numerous government agencies, showcasing the broad impact of such incidents. Furthermore, the consequences of supply chain attacks extend beyond operational disruptions, posing serious threats to public trust, privacy, and national security. When malicious software is distributed through legitimate channels, even well-prepared organizations may struggle to detect and mitigate it. This highlights the importance of software vulnerability management and the need for innovative solutions that enhance transparency and enable proactive defense against emerging risks.

2 Literature Review

In response to these challenges, the Biden administration issued Executive Order 14028 on Improving the Nation's Cybersecurity in 2021, mandating the submission of Software Bill of Materials (SBOM) for software products purchased by federal agencies [2, 3]. This initiative is part of a larger shift to strengthen accountability and security in the software supply chain. As of 2024, regulations and standards related to SBOM are actively being implemented in various industries, including healthcare, automotive, and IoT devices, primarily in the U.S., Europe, and Japan [4]. These regulations aim to help identify and mitigate vulnerabilities by providing a clear understanding of the components of software systems.

SBOM provides a detailed list of the libraries, dependencies, and other components used in software applications [5]. It is similar to an ingredient list on food packaging, helping stakeholders identify potentially harmful components, assess risks, and make informed decisions about software usage. This is particularly important in today's software development environment, where code reuse and open-source libraries are prevalent. According to Synopsys' 2021 study, 98% of analyzed codebases rely on open-source software, and 84% contain at least one known security vulnerability [6]. If these vulnerabilities are not addressed, they can create systemic risks across the software ecosystem.

While SBOM provides valuable information, it is insufficient on its own for vulnerability management. Simply identifying vulnerabilities does not enable organizations to assess the actual risk or prioritize them. To address this, the Cybersecurity and Infrastructure Security Agency (CISA) and the National Telecommunications and Information Administration (NTIA) introduced the concept of Vulnerability Exploitability eXchange (VEX). VEX documents complement SBOM by providing a detailed assessment of whether identified vulnerabilities in specific software environments are actually exploitable. By clarifying the exploitability status, VEX documents help organizations prioritize critical vulnerabilities and allocate resources effectively [7]. Gartner's Leader's Guide to Software Supply Chain Security emphasizes the need for continuous risk management using SBOM and VEX [8].

However, implementing VEX presents significant challenges. Evaluating the exploitability of vulnerabilities requires high-level analysis, which is time-consuming and resource intensive. The current approach to VEX generation lacks the automation necessary for widespread adoption, making it ineffective in rapidly changing software environments. Additionally, the technology needed to verify whether software component vulnerabilities are truly exploitable is not yet sufficiently developed. Addressing these limitations is crucial for advancing vulnerability management and strengthening software supply chain security.

This study proposes a new framework for automating the generation of SBOM and VEX documents. This approach aims to simplify and improve the accuracy of vulnerability identification and assessment, reducing the time and effort required for vulnerability management and enhancing the reliability of exploitability assessments. Through this, organizations can respond to threats more quickly, prioritize critical vulnerabilities, and strengthen their overall cybersecurity posture. Furthermore, by integrating the automation of SBOM and VEX generation into the vulnerability management process, this approach helps manage the complexities of the modern software ecosystem and ultimately contributes to ensuring a safer digital environment.

3 Methodology

3.1 Utilizing LanGraph to Improve LLM Model

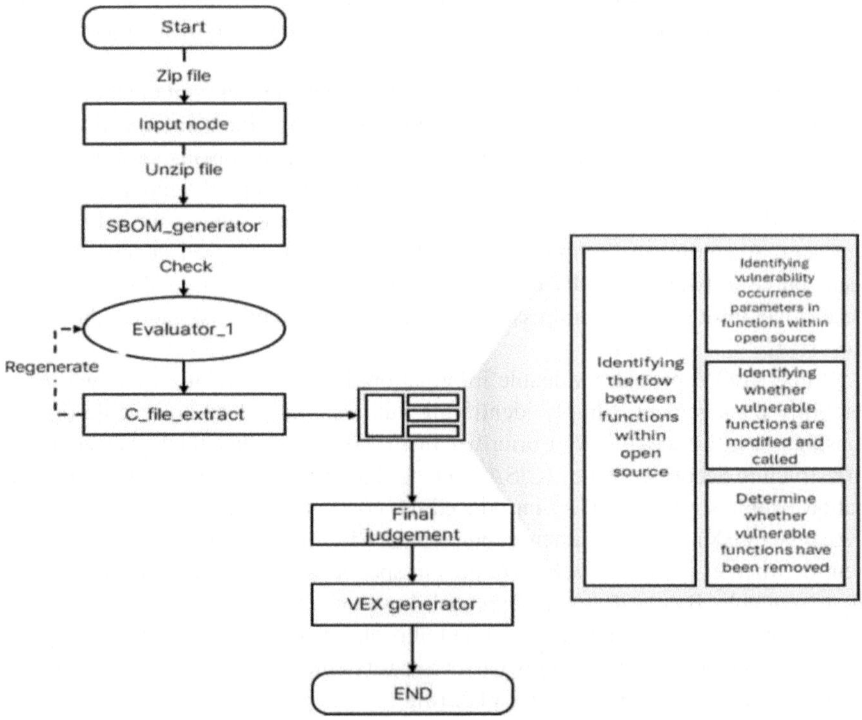

Fig. 1. "VEXine" LangGraph

This study proposes "VEXine", a technique for automating the generation of SBOM (Software Bill of Materials) and VEX (Vulnerability Exploitability eXchange) documents using Transformer-based LLM models and LangGraph. This technology supports identifying vulnerabilities, assessing their exploitability, and enhancing software security. Additionally, it introduces new criteria for evaluating vulnerability statuses in VEX documents. To develop a model for assessing vulnerabilities through SBOM and VEX document generation, we introduce LangGraph, a framework for designing complex workflows and applications using LLMs. LangGraph allows users to provide source code to an LLM and design operational logic as a decision-based graph. It generates nodes and edges based on functionality and integrates multiple agents to ensure stability within software development environments. Figure 1 illustrates a graph connecting nodes to optimize the analysis of vulnerabilities.

Table 1. "VEXine" LangGraph GraphState

Definition of Graph State	Explanation	Status
Input_path	The Zip file path entered by the user	Str
Node_path	The path being executed by the node	Str
Total_time	Total execution time	Int
Time_spent	Time taken per node	Int
Status	Node progress status	Int
Output_path	File path for saving execution results	str

LangGraph includes GraphState, a mechanism for sharing state information, such as current nodes and data, during workflow execution. Table 1 shows GraphState sharing various types of information, including input paths and execution progress. For example, after a ZIP file input is processed, the state updates to "ZIP file extracted." Similarly, successful SBOM generation updates the state, enabling dynamic execution of subsequent nodes based on shared states. To overcome LLM limitations like hallucination and false positives, we introduce an Evaluator and employ multi-stage learning with iterative review. Cross-validation allows a secondary LLM to refine initial outputs, enhancing overall response quality.

3.2 Detection Methodology Area

This study, the complex issue of vulnerability manifestation is subdivided into four subproblems. Each problem represents an independent task from the perspective of vulnerability analysis, which can be defined as follows:

Detection of Vulnerabilities Based on Source Code Flow. Software vulnerabilities are often influenced by the data flow and execution paths within the source code. In particular, the transfer of data between functions and alterations in execution flow can lead to security issues. For instance, vulnerabilities may arise when functions handling sensitive data are invoked out of order, or when necessary, security validation processes are omitted.

Determination of Whether Vulnerable Functions Have Been Removed. If vulnerable functions identified through CVE information Remain unremoved or persist with security vulnerabilities, further analysis is required to assess their state and potential risks. This study evaluates the security vulnerabilities of such functions by applying three subsequent analytical steps when the vulnerable functions remain unremoved.

Detection of Modification and Invocation of Vulnerable Functions. Vulnerable functions serve as critical security routines within a system; however, If the code is altered or abnormal invocations occur, these security routines may fail to operate correctly. For example, if a conditional statement preceding the invocation of an authentication function is altered to bypass authentication, security can be compromised. This study leverages CVE information to determine whether functions with known vulnerabilities are actually invoked in the source code. The analysis involves abstraction and hashing of variables and parameters in the source code to detect any modifications to these functions.

Detection of Vulnerable Function Calls via User Input Parameters. User input is one of the primary causes of software vulnerabilities. When externally provided input values are passed to sensitive functions without proper validation, attackers can exploit these inputs. For example, if user input is directly passed to a dynamic code execution function or a database query generation function, vulnerabilities such as Remote Code Execution (RCE) or SQL injection may arise (Fig. 2).

```
def execute_user_input(request):
user_input = request.GET["input"]
exec(user_input)
```

Fig. 2. Examples of vulnerabilities in user input parameters

In the above code, request. GET["input"] takes the value provided by the user and directly passes it to the exec() function for execution. This creates a pathway for attackers to insert malicious code and compromise the system. As demonstrated in this example, to determine whether parameters receiving user input actually reach a vulnerable function, parameters are categorized based on the type of Attack Surface [12], and it is further analyzing whether they reach the vulnerable function.

3.3 Definition of Detectable Area

Detection of Vulnerabilities Based on Source Code Flow. This approach analyzes the flow between functions in the source code used in open source to determine whether vulnerabilities are manifested.
Dataset Collection and Preprocessing: The dataset used for model training includes approximately 180,000 data samples from the ICSE '24 paper, "Dataflow Analysis-Inspired Deep Learning for Efficient Vulnerability Detection" [13]. To understand the flow of the source code, the static analysis tool 'Joern' is used to generate CFG (Control

Flow Graph) and AST (Abstract Syntax Tree), which are then converted into graph formats for preprocessing. After preprocessing, the abstract source code is examined for vulnerable API calls, variables, data types, literal values, operators, and other factors contributing to potential vulnerabilities. Following the embedding and bit-vectorization process, the data is prepared for model training. The trained model predicts the likelihood of vulnerability manifestation.

Model Components: To analyze function flow, the GGNN-based 'DeepDFA' model for source code flow detection and the 'LineVul' model, designed for context interpretation, were designed and trained [14].

Embedding Layer: The embedding layer leverages pre-trained weights from a Roberta-based encoder to generate dense vector representations of input sequences, capturing contextual and semantic information essential for vulnerability classification.

Graph Neural Network (FlowGNN): The FlowGNN module processes graph-based features derived from the Code Property Graph (CPG), emphasizing control and data flow. It captures relationships among critical code elements necessary for vulnerability detection.

Concatenation Layer: This layer merges the embeddings generated by the Roberta encoder and FlowGNN, integrating sematic and structural information. It provides a unified representation for Downstream classification tasks.

Classification Head: The classification head determines whether a given source code snippet is vulnerable. Its structure includes a Dense layer that combines and reduces the dimensionality of the embeddings, a Dropout layer that regularizes the model to prevent overfitting, and an Output layer that classifies the features into two logical categories(vulnerable or non-vulnerable.)

Loss Function: The binary classification model is optimized using a cross-entropy loss function, as defined in (1). This function measures the difference between the actual label of the i-th sample y_i and the predicted probability $\widehat{y_i}$ for the same sample.

$$L = -\frac{1}{N}\sum_{i=1}^{N}[y_i log(\widehat{y_i}) + (1 - y_i) log(1 - \widehat{y_i})] \tag{1}$$

Training Procedure: The model is trained end-to-end using the AdamW optimizer and a linear warm-up scheduler. To mitigate overfitting and ensure numerical stability, dropout regularization and gradient clipping are employed.

Determining the Removal of Vulnerability Manifestation Functions. This approach evaluates whether functions manifesting vulnerabilities within the source code of open source projects actually exist. Similar to the analysis process in Sect. 3, "Analyzing the Modification and Invocation of Vulnerable Functions," this method compares existing functions with the current code through abstraction and hashing of the source code to determine whether the vulnerable function has been removed. If the vulnerable function exists, the status is marked as 'Affected'; if it has been removed, the status is marked as "Not Affected." When the vulnerable function remains, the previously defined three analytical steps are applied to evaluate its security vulnerabilities.

Analysis of Vulnerable Function Modification and Invocation. This method examines whether functions manifesting vulnerabilities within the source code of open source projects are actually invoked. Using information about open source code and its constituent functions, as verified through SBOM, this method abstracts and hashes variables, parameters, and other elements within the source code where CVEs occur to detect matches in function names and contents.

During the detection of matches, the Backward Slicing technique is applied to isolate only the code that influences a specific function name. This approach allows effective detection even if users arbitrarily modify functions outside of vulnerable ones. Once a vulnerable function is identified, the Joern analysis tool is used to preprocess and abstract the source code. Based on the function name, the function's structure is analyzed to verify whether it is actually invoked. If the vulnerable function is invoked, the status is marked as 'Affected'; if not, the status is marked as "Not Affected." This process effectively detects modified functions or abnormal invocation states.

Analysis of Vulnerable Manifestation Parameters Within Functions. This approach determines whether parameters that accept user input in open source code are passed to vulnerable functions identified from CVE information. The Attack Surface of a parameter can be broadly classified into three categories: Entry Point, Target, and Mechanism [14]. In this methodology, the scope is narrowed to Entry Point to identify parts directly related to user input.

The Entry Point of a parameter is further classified into System, Code, Program, and Network. To detect and analyze all scenarios within each category, parameters are identified based on representative functions associated with each type.

Table 2. Function Types Categorized by Entry Points

Attack Surface	Function
System	malloc, calloc, realloc
Code	scanf, gets, fscanf, fgets, read
Network	packet, ip, address, socket, tcp, udp, net, data
Program	fd & read, write, open, close

Steps for Identifying Vulnerable Parameters is as follows.

Identify Entry Point Functions: The first step involves examining the code for functions listed in Table 2 within the code, focusing on Entry Points. If any of the functions in Table 2 are present in the code, the subsequent analysis process is carried out.

Identify Parameters in the Identified Function Declarations: If a function matches one in Table 2, the function declaration is examined to identify its parameters. Since the position of user-related input parameters may vary depending on the function, It is essential to identify and analyze the parameters related to user input.

Analyze the Flow of Identified Parameters: The parameters identified in step 2 are targeted, and the Joern tool is used to trace all instances where the parameter is referenced. The Joern tool generates a Code Property Graph (CPG), making it highly effective for understanding precise code flows. Thus, it becomes possible to identify all parts of the code where the parameter values are used.

Analyze Flow Results: Affected/Not Affected: By referencing the "func_name" column in the CVE DB, it is determined whether the vulnerable functions exist at any of the locations where the parameter values are used. If a vulnerable function is found, the result is returned as "Affected." This indicates that the parameter containing the user input is also used in a vulnerable function, meaning that an attacker's input could reach the vulnerable function. By identifying the Attack Surface of parameters and tracing parameter values based on specific functions and code flows, it becomes possible to determine whether the parameter values reach a vulnerable function. This process is significant because it enables the identification of parts of the code directly affected by user input.

3.4 VEXine Final Evaluation Methodology

Advanced Vulnerability Status Evaluation Criteria

Table 3. Methodology for Assessing Project Vulnerability Status

Flow analysis	Functions removed	Functions called	Does input affect parameters?	Vulnerability-exposing parameter	Final result
Recheck	Not Affected				Not Affected
	Affected	Not Affected		Affected	Not Affected
				Not Affected	Not Affected
		Affected	Yes (Affected)	Not Affected	Not Affected
		Affected	No (Not Affected)	Not Affected	Affected
		Affected		Affected	Affected

Final Vulnerability Status Evaluation Method: Leveraging the VEX classification methodology outlined above, vulnerabilities are assessed for each definition, enabling a comprehensive evaluation of the vulnerability status of the project. The methodology for evaluating the project's vulnerability status is detailed in Table 3. As illustrated in Table 3, the results of each methodology can be analyzed using a Decision Tree algorithm to determine the final structural impact.

Determining the Removal of Vulnerability Manifestation Functions: If a vulnerability manifestation function has been removed, it is recognized as no longer existing in the project, and the final status is marked as "Not Affected". Conversely, if the function remains and has not been removed, the final result depends on the outcomes of the

"Vulnerability Manifestation Parameter Analysis" and "Modification and Invocation of Vulnerable Functions".

Comparison of Results from "Vulnerability Manifestation Parameter" and Vulnerable Function Modification and Invocation Analyses": If the vulnerable function is not invoked within the project, the function is deemed to have no potential for vulnerability manifestation, even if vulnerable parameters exist within the function. In such cases, the final status is marked as "Not Affected", regardless of the results from the "Vulnerability Manifestation Parameter Analysis."

However, if the vulnerable function is invoked, the function's parameters are analyzed to assess whether they are associated with user input. If the function's parameters are unrelated to user input, it is determined that the function itself inherently possesses vulnerabilities, and the final status is "Affected". On the other hand, if the function's parameters are related to user input, the analysis determines whether the vulnerable parameter receives input from external sources. If the parameter both receives external input and is influenced by the input, the final status is marked as "Affected". This process of analyzing and comparing the results of vulnerability manifestation allows for determining the final status.

Evaluation Criteria Based on the Time of Generation. Based on the final vulnerability evaluation results outlined above, the vulnerability status evaluation results can be presented as a VEX specification that reflects the criteria set by CISA for VEX vulnerability status evaluation. The results of the VEX document generated by "VEXine" enhances existing VEX vulnerability status evaluation criteria by incorporating the advanced evaluation criteria by incorporating the advanced evaluation criteria and generation timing defined by "VEXine." Table 4 illustrates the vulnerability status evaluation criteria used by "VEXine." [15].

Table 4. Vulnerability Status Evaluation Criteria for "VEXine"

Types of judgment	Explanation	Creation time
Affected	State with potential for exploitation, recommended to implement measures for modification or mitigation	At the time of initial VEX creation
Not affected	State with no associated risk related to the vulnerability, no modification required	
Fixed	After the assessment of 'affected', the vulnerability has been resolved through a patch or update (Affected → Not affected)	Created during reinspection
Under investigation	After the assessment of 'affected', the vulnerability remains unresolved (Affected → Affected)	

During the initial vulnerability impact assessment for a software product and generating a VEX document, the analysis results are categorized as either "Affected" or "Not Affected". For subsequent vulnerability impact assessments of the same product, the previously generated VEX document is referenced. Components marked as "Affected" are reanalyzed to Verify whether modifications have resolved the vulnerabilities or if they persist. The results of the analysis are categorized as "Fixed" or "Under Investigation":

Fixed: Indicates that the vulnerability has been resolved that the vulnerability has been resolved, and the component is now "Not Affected".

Under Investigation: Returned when reanalysis shows that the vulnerability has not been resolved, and the component remains "Affected".

4 Data Collection and Study Area Section

4.1 Setting up a CVE-Based Database Using LLM

"VEXine" generates VEX documents by leveraging CVE (Common Vulnerabilities and Exposures) information. This tool extracts details such as the names of code files, function names, and function contents where vulnerabilities occur in open-source code, based on the CVE descriptions in the CVE data [9, 10].

By collecting commit information from projects where vulnerabilities have been identified and patched, a comprehensive CVE database was constructed. Additionally, irregular CVE data were reformatted and restructured into a detailed database using LLM. The collected information served as input for the LLM T5 model, which utilized its text generation capabilities through specific prompts. To further enhance domain specific tasks, the T5 model was fine-tuned using the LoRA (Low-Rank Adaptation) technique. During this finetuning process, descriptions from NVD [11] were used as input, while the extracted key information was designated as labels. Supplementary information was also collected using Python web scraping, enabling the construction of a CVE database, as demonstrated in Table 5. However, a critical problem Analysis Despite these advancements, CVE descriptions have inherent limitations that restrict their practical use in vulnerability management and analytical data collection:

Inconsistencies in CVE Descriptions: CVE descriptions are often unstructured and vary significantly depending on the author's style and explanation methods.

Lack of Structure: Technical information is provided solely in textual form, making it difficult to extract critical details such as affected system environments, vulnerability types, and severity levels using standard regular expression parsers.

Ambiguity in Descriptions: Variations in expressions for the same vulnerability can result in data duplication or ambiguous interpretations, causing confusion.

Additional Efforts Required by Analysts: Analysts frequently need to perform additional searches or integrate supplementary information to obtain the necessary data.

Insufficient Technical Context: Most CVE descriptions merely list the technical issues where vulnerabilities were discovered, lacking essential details such as Proof-of-Concept (POC) code, exploit scenarios, specific component or library function names, and execution scenarios in real environments.

Challenges in Structuring and Normalization: Structuring and normalizing CVE data heavily rely on manual efforts or limited automation tools, making it difficult to process large volumes of CVE data in realtime.

Variations in how authors describe the same vulnerability differently, it becomes challenging to eliminate duplicate data and conduct integrated analysis, often resulting in confusion. Furthermore, missing details about the scope of impact in real world environments (e.g., whether exploitation is possible under specific conditions) and the absence of specific component and function names pose additional challenges. These details are critical for developers or security managers to ensure immediate remediation and assess patch status.

The CVE-Description Database (hereafter referred to as CVE-DB) is designed to include information such as that shown in Table 5. While details like affected versions and CVSS scores are also included in existing CVE-Details pages, the CVE-DB goes further by standardizing and organizing this date.

Table 5. Information about CVE included in the Database

Index	Index number	Vendor	Document Author
CVE-ID	CVE-ID Number	**Affected_Product**	Affected Environment and Version/Product
CWE-ID	CWE-ID Number	**Affected_Version**	Affected Library Version
Description	Vulnerability	**File_name**	Vulnerable File Name
Impact-C	Confidentiality	**Func_name**	Vulnerable Function Name
Impact-I	Integrity	**Old**	Problematic Code
Impact-A	Availability	**New**	Patched Code
Exploitability	Technical Complexity (Low, High)	**Severity**	CVSS Score
Published_Date	Initial Publication Date	**Expected_Vulnerability**	Expected Impact of the Vulnerability
Last_Modified_Date	Last Modified Date	**Attack_Vendor**	Network, Physical and so on
Privileges_Required	Privilege Requirement	**Vulnerability_type**	Attack Type

Although the CVE-DB provides additional related content, it still faces limitations in collecting all the essential information.

5 Result

5.1 Model Training and Evaluation

After completing the training, the trained model was loaded, and its predictions showed an accuracy of 96.03% and an F1-score of 94%. Based on the training results, the likelihood of vulnerability manifestation in the source code is output as a probability. Predictions are presented as, the results are expressed as "0: Not Affected" or "1: Affected." (Table 6).

Table 6. Performance metrics for DeepDFA and LineVul training

	precision	recall	f1-score	support
0	0.99	0.99	0.99	9244
1	0.93	0.83	0.88	768
accuracy			0.98	10012
macro avg	0.96	0.91	0.94	10012
micro avg	0.98	0.98	0.98	10012

5.2 Analysis Time for Large-Scale and Small-Scale Projects

The analysis of large-scale projects took up to 40 min, while small-scale projects were completed in as little as 3 min. These results demonstrate VEXine's flexibility and efficiency in handling projects of varying sizes.

5.3 Validation of VEXine Detection Accuracy

Based on validation conducted on the top 100 GitHub repositories, "VEXine" successfully detected 341 out of 352 CVEs, achieving a True Positive rate of 97%. The False Negative rate was confirmed to be 2.7%, and the False Positive rate was 0.3%. These results demonstrate the high reliability of VEXine's Affected/Not Affected classification and its potential for effective application in real-world projects.

6 Conclusion

"VEXine" is an innovative tool designed to enhance supply chain security by systematically analyzing software vulnerabilities and supporting organizations in building sustainable and trustworthy systems. By identifying open-source components within source code, generating SBOMs, and evaluating CVEs, "VEXine" determines the potential manifestation of vulnerabilities and produces comprehensive VEX documents. Leveraging advanced methodologies, such as LangGraph-based workflow design and the GraphState mechanism, it would automate the end-to-end process from SBOM creation to VEX generation, simplifying the complex process of vulnerability analysis for developers and security managers. Through multifaceted evaluation approaches, "VEXine" establishes itself as a reliable tool that assesses vulnerabilities from multiple dimensions while introducing advanced criteria for incorporating analysis time and status into its assessments. This approach positions "VEXine" as a potential standard for software supply chain security, offering scalable solutions across industries and opening new horizons for secure software development and maintenance.

Despite its contribution, this study has limitations such as atypical characteristics of CVE data and potential false positive rate in detecting vulnerabilities. To solve these problems, further research and policies on CVE data standardization and model optimization are needed. In addition, for the application of "VEXine" to various industrial environments such as OT, IoT, automotive, and medical care, the focus should be on expanding programming languages and securing various CVE datasets.

References

1. Check Point: Protect yourself against the Apache Log4j vulnerability (2021)
2. Federal Register: Public listening sessions on advancing SBOM technology, processes, and practices, 1 June 2022
3. Choi, Y.: Trends in U.S. software supply chain security policy: Focusing on SBOM cases. J. Inf. Secur. **32**(5), 7–14 (2022)
4. U.S. Food and Drug Administration (FDA): Cybersecurity in medical devices: Quality system considerations and content of premarket submissions draft guidance for industry and Food and Drug Administration staff, 8 April 2022
5. Sohn, H., Kim, D., Kim, S.: A study on software supply chain security policy for enhancing cybersecurity: focusing on SBOM policy cases. J. Digit. Converg. **20**(2), 9–20 (2022)
6. Synopsis: New Synopsys report finds 74% of codebases contained high-risk open-source vulnerabilities, surging 54% since last year, 27 February 2024. https://investor.synopsys.com/news/news-details/2024/New-Synopsys-Report-Finds-74-of-Codebases-Contained-High-Risk-Open-Source-Vulnerabilities-Surging-54-Since-Last-Year/default.aspx
7. Park, C.: Enhancing the effectiveness of the Software Bill of Materials (SBOM) with VEX, Center for Software Security and Assurance (CSSA) (2022). https://www.dbpia.co.kr/journal/articleDetail?nodeId=NODE11149583
8. Gartner: Innovation insight for SBOMs, 14 February 2022
9. Fan, J., Li, Y., Wang, S., Nguyen, T.N.: A C/C++ code vulnerability dataset with code changes and CVE summaries. In: MSR 2020: The 17th International Conference on Mining Software Repositories, pp. 1–5. ACM (2020). https://doi.org/10.1145/3379597.3387501
10. GitHub, ZeoVan/MSR_20_Code_vulnerability_CSV_Dataset (n.d.). https://github.com/ZeoVan/MSR_20_Code_vulnerability_CSV_Dataset
11. National Vulnerability Database (NVD). https://nvd.nist.gov/
12. Doe, J., Smith, A.: A grounded theory-based approach to characterize software attack surfaces. arXiv (2021). https://arxiv.org/abs/2112.01635
13. Steenhoek, B., Gao, H., Le, W.: Dataflow analysis-inspired deep learning for efficient vulnerability detection. In: 2024 IEEE/ACM 46th International Conference on Software Engineering (ICSE 2024), pp. 1–13. ACM (2024)
14. GitHub, ISU-PAAL/DeepDFA (n.d.). https://github.com/ISU-PAAL/DeepDFA
15. Cybersecurity and Infrastructure Security Agency (CISA), Vulnerability Exploitability eXchange (VEX) – Use Cases (2022). https://www.cisa.gov/resources-tools/resources/vulnerability-exploitability-exchange-vex-use-case-document-april-2022

Enhancing Link Prediction with Fuzzy Graph Attention Networks and Dynamic Negative Sampling

Jinming Xing[1(✉)], Ruilin Xing[2], Chang Xue[3], and Dongwen Luo[4]

[1] North Carolina State University, Raleigh, NC, USA
jxing6@ncsu.edu
[2] Guangxi University, Nanning, Guangxi, China
[3] Yeshiva University, New York, NY, USA
cxue@mail.yu.edu
[4] South China University of Technology, Shenzhen, Guangdong, China

Abstract. Link prediction is crucial for understanding complex networks but traditional Graph Neural Networks (GNNs) often rely on random negative sampling, leading to suboptimal performance. This paper introduces Fuzzy Graph Attention Networks (FGAT), a novel approach integrating fuzzy rough sets for dynamic negative sampling and enhanced node feature aggregation. Fuzzy Negative Sampling (FNS) systematically selects high-quality negative edges based on fuzzy similarities, improving training efficiency. FGAT layer incorporates fuzzy rough set principles, enabling robust and discriminative node representations. Experiments on two research collaboration networks demonstrate FGAT's superior link prediction accuracy, outperforming state-of-the-art baselines by leveraging the power of fuzzy rough sets for effective negative sampling and node feature learning.

Keywords: Link Prediction · Graph Neural Networks · Fuzzy Rough Sets · Negative Sampling

1 Introduction

Link prediction has emerged as a crucial task in network analysis with extensive applications across diverse domains. In medical sciences, it aids in predicting protein-protein interactions and drug-target associations; in financial systems, it helps detect fraudulent transactions and assess credit risks; and in chemistry, it facilitates the discovery of novel molecular structures and chemical reactions. The ability to accurately predict potential connections in these complex networks has significant implications for scientific advancement and practical applications.

Graph Neural Networks (GNNs) have demonstrated remarkable success in link prediction tasks, primarily due to their inherent capability to capture and process structural information in graph-structured data. However, a critical limitation in existing GNN-based approaches lies in their negative sampling methodology. Contemporary methods typically employ random sampling strategies to

A. Ito (Ed.): ICMSSP 2025, LNNS 1637, pp. 187–197, 2025.
https://doi.org/10.1007/978-3-032-05994-9_16

select negative edges, disregarding the rich semantic and structural information encoded in node representations. This oversight significantly hampers the training process, resulting in slower convergence rates and suboptimal model performance. An ideal negative sampling mechanism should not only leverage node embeddings effectively but also adaptively select high-quality negative samples based on the model's current state, ensuring both dynamic responsiveness and sampling accuracy.

While various methodologies have been explored to enhance link prediction accuracy, the potential of fuzzy rough sets-a mathematical framework for measuring fuzzy relations and handling uncertainty-remains largely unexplored in the context of GNNs and link prediction. This theoretical framework offers unique advantages in capturing imprecise relationships and handling ambiguous data structures, making it particularly suitable for network analysis tasks.

To address these limitations and leverage the untapped potential of fuzzy rough sets, we propose a novel fuzzy rough sets-based negative sampling strategy called Fuzzy Negative Sampling (FNS). This approach systematically evaluates candidate negative edges through their fuzzy lower approximation values, selecting the top K candidates as negative training instances. Furthermore, we introduce Fuzzy Graph Attention Network (FGAT), an enhanced graph neural architecture designed to aggregate neighboring node information in a more robust and effective manner.

The main contributions of this work can be summarized as follows:

- We introduce FNS, a novel negative sampling framework that leverages fuzzy rough sets theory to identify high-quality negative edges, significantly improving the effectiveness of the training process in link prediction tasks.
- We propose FGAT, an innovative graph attention network that incorporates fuzzy rough set principles to achieve more robust and discriminative node representations.
- We conduct comprehensive experiments across two real-world datasets, demonstrating the effectiveness of our proposed framework.

2 Related Work

Graph Neural Networks have demonstrated remarkable versatility across various graph-based learning tasks. In node classification, seminal works like GraphSAGE [1] and Graph Attention Networks (GAT) [2] have established foundational approaches for learning node representations through neighborhood aggregation. GCN [3] introduced convolutional operations on graphs, enabling efficient feature propagation across network structures. For graph classification tasks, hierarchical pooling mechanisms have been developed, with DiffPool [4] and TopKPool [5] proposing learnable strategies to generate graph-level representations. In the context of link prediction, SEAL [6] pioneered the use of local subgraphs for edge existence prediction, while VGAE [7] employed variational autoencoders for learning edge formation patterns. Recent advances include NetSight [8], which introduces a novel graph attention framework to capture both global and local spatial-temporal co-dependencies for traffic forecasting.

Fuzzy rough sets theory, initially proposed by Dubois and Prade [9], has evolved into a powerful framework for handling uncertainty and imprecision in data analysis. In feature selection, Jensen and Shen [10] developed fuzzy-rough attribute reduction algorithms that significantly outperform traditional approaches in identifying relevant features while maintaining information fidelity. The application of fuzzy rough sets in medical diagnosis has been exemplified by works such as [11], where they effectively handle the inherent uncertainty in patient data for more accurate disease classification. For uncertainty measurement, the framework has been extensively studied in theoretical works by [12,13], establishing mathematical foundations for quantifying various types of uncertainty in data relationships. Recent developments include hybrid approaches combining fuzzy rough sets with deep learning and applications in big data analytics [14], demonstrating the framework's adaptability to modern computational challenges.

Except the innovative applications of Graph Neural Networks and fuzzy rough set theory in handling complex data and multi-task learning, against the backdrop of rapid advancements in machine learning, applications like knowledge graph technology in intelligent question-answering systems have become a key area of development for graph learning methods, effectively integrating multiple data sources to support flexible knowledge processing [15]. Similarly, multi-model integration technology applied to automated generation systems has significantly enhanced content generation flexibility and quality, providing valuable insights for representation learning based on graph data [16]. In network security, Graph Neural Networks have demonstrated strong generalization capabilities in supporting Botnet detection through machine learning, precisely identifying abnormal behaviors and enhancing network defense levels [17].

In recent years, the scope of machine learning applications has continuously expanded [18–23], especially in network optimization [24], image processing, and VR fields. Blockchain-enhanced image retrieval systems, for example, have brought revolutionary improvements in data security and retrieval efficiency [25]. In VR and robotic interaction, AI-vision-powered intelligent systems have explored new methods for balancing real-world interaction and virtual immersion, making human-computer interaction more natural and seamless [26]. Furthermore, the integration of fuzzy rough set theory with deep learning has also achieved breakthroughs in traffic data prediction, enabling more accurate short-term forecasting through multi-source data fusion, effectively adapting to the dynamic demands of complex data environments [27]. Overall, these technological innovations not only showcase the broad applicability of graph learning methods and fuzzy rough sets across different task scenarios but also reinforce their theoretical and practical value in big data and intelligent applications.

3 Methodology

Figure 1 illustrates our proposed framework, which consists of two main components:

- **Fuzzy Negative Sampling:** A mechanism that selects high-quality negative edges based on fuzzy similarities, where negative edges with high fuzzy similarity are dynamically selected for the FGAT framework's training.
- **FGAT Convolution Layer:** A specially designed layer for effective neighbor node information aggregation. Multiple FGAT convolution layers are stacked to capture multi-hop information.

In the following sections, we first detail the computation of fuzzy similarities using fuzzy rough sets for high-quality negative edge selection. Subsequently, we elaborate on the FGAT architecture, followed by a comprehensive framework summary.

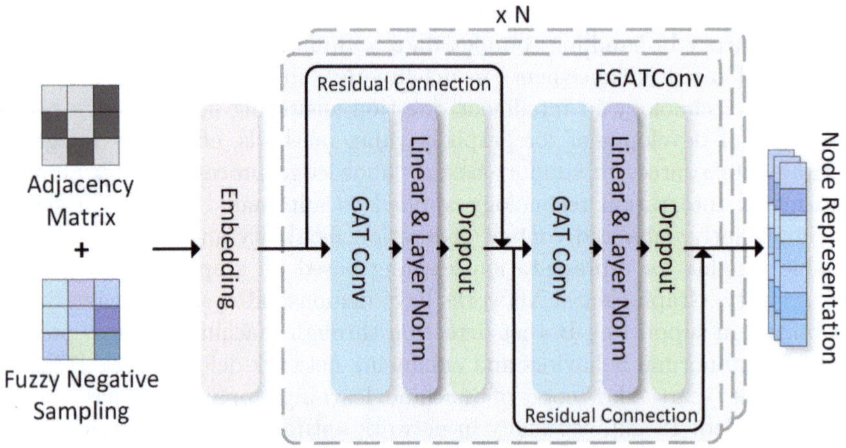

Fig. 1. The FGAT Framework

3.1 Fuzzy Negative Sampling

A fuzzy information system is defined as a tuple (U, A, V, f), where U represents a non-empty finite set of samples, A denotes the finite set of sample attributes, V represents the domain of all attributes in A, expressed as $V = \bigcup_i V_i$ where V_i is the domain of attribute i, and f is a mapping function $U \times A \rightarrow V$ [11].

For an attribute set $B \subseteq A$ and a fuzzy equivalence relation R, we can compute a coverage of the universe U. For a sample x, we denote its coverage under the fuzzy equivalence relation R as $[x]_R$. The membership of a sample y to the coverage $[x]_R$ is defined as $[x]_R(y) = R(x, y)$, where $R(x, y)$ quantifies the similarity between samples x and y under relation R. For any sample $x \in U$ and subset $X \subseteq A$, the fuzzy lower and upper approximations of sample x to X are defined as [11]:

$$\begin{aligned} \underline{R_S}X(x) &= \inf_{y \in U} S(N(R(x,y)), X(y)), \\ \overline{R_T}X(x) &= \sup_{y \in U} T(R(x,y), X(y)) \end{aligned} \tag{1}$$

where S and T represent fuzzy triangular conorm (S-norm) and fuzzy triangular norm (T-norm) respectively, and $N(x) = 1 - x$.

Using the conventional min-max version of T and S norms, for a set of samples d_i of class i and corresponding attribute set $B \subseteq A$, Eq. 1 can be reformulated as:

$$\begin{aligned} \underline{R_B}d_i(x) &= \inf_{y \in U} \max(1 - R(x,y), d_i(y)), \\ \overline{R_B}d_i(x) &= \sup_{y \in U} \min(R(x,y), d_i(y)) \end{aligned} \tag{2}$$

To capture non-linear high-level similarities, R typically employs kernel functions, including the Gaussian kernel: $k_G(x,y) = \exp(-\frac{||x-y||^2}{\delta})$, exponential kernel: $k_E(x,y) = \exp(-\frac{||x-y||}{\delta})$, and rational quadratic kernel: $k_R(x,y) = 1 - \frac{||x-y||^2}{||x-y||^2+\delta}$.

During each training epoch, negative links are dynamically selected based on their quality scores. For any potential negative link with end nodes (x, y), the quality score is computed as:

$$Score(x,y) = \alpha \times \underline{R_B}d_y(x) + (1-\alpha) \times \underline{R_B}d_x(y) \tag{3}$$

where α is a hyperparameter.

While computing quality scores for all possible negative edges and selecting the top k would be optimal, this approach becomes computationally intractable for large dense graphs. For a graph with N nodes and E edges, there exist $N \times (N-1) - E$ potential directed negative edges. To address this computational challenge, we randomly select $2E$ negative edges and select the top E edges among them. This strategy reduces the computational complexity from $N \times (N-1) - E$ to $2E$ while maintaining near-optimal performance.

The selected top E negative edges are combined with the original positive edges to form the training dataset. To prevent class imbalance issues, we maintain an equal number of selected negative edges and original positive edges.

3.2 FGAT Convolution Layer

The FGAT convolution layer integrates GAT convolution layers with linear layers, incorporating layer normalization for training acceleration and dropout mechanisms for effective regularization.

Given an undirected graph $G = (V, E)$, where V represents the set of nodes and E denotes the set of edges, each node $v \in V$ is associated with a feature vector $\mathbf{h}_v \in \mathbb{R}^F$, where F represents the dimension of input features per node. The

FGAT layer aims to compute updated node representations $\mathbf{h}'_v \in \mathbb{R}^{F'}$, where F' denotes the output feature dimension, by performing weighted aggregation of features from each node's neighborhood.

For a node pair consisting of node v and its neighbor u, the attention coefficient e_{vu} is computed through:

$$e_{vu} = \text{LeakyReLU}\left(\mathbf{a}^T \left[\mathbf{W}\mathbf{h}_v \parallel \mathbf{W}\mathbf{h}_u\right]\right) \tag{4}$$

where:

- $\mathbf{W} \in \mathbb{R}^{F' \times F}$ represents a learnable weight matrix that transforms node features linearly.
- $\parallel$ indicates vector concatenation.
- $\mathbf{a} \in \mathbb{R}^{2F'}$ denotes a learnable weight vector.
- LeakyReLU serves as the activation function, typically configured with a small negative slope (e.g., 0.2).

The attention coefficients then undergo normalization across each node's neighborhood using the softmax function:

$$\alpha_{vu} = \frac{\exp(e_{vu})}{\sum_{k \in \mathcal{N}(v)} \exp(e_{vk})} \tag{5}$$

where $\mathcal{N}(v)$ represents the neighborhood set of node v.

The normalized attention scores α_{vu} facilitate the computation of updated node features $\mathbf{h}'_v$ through weighted aggregation:

$$\mathbf{h}'_v = \sigma\left(\sum_{u \in \mathcal{N}(v)} \alpha_{vu} \mathbf{W}\mathbf{h}_u\right) \tag{6}$$

where σ represents a non-linear activation function, typically implemented as ReLU.

To enhance model robustness and representational capacity, the GAT layers employ multi-head attention mechanisms. Specifically, K independent attention heads operate in parallel, each generating distinct attention coefficients and feature representations. These representations are subsequently concatenated to produce the final output:

$$\mathbf{h}'_v = \parallel_{k=1}^{K} \sigma\left(\sum_{u \in \mathcal{N}(v)} \alpha_{vu}^{(k)} \mathbf{W}^{(k)} \mathbf{h}_u\right) \tag{7}$$

where $\alpha_{vu}^{(k)}$ and $\mathbf{W}^{(k)}$ correspond to the attention coefficient and weight matrix of the k-th attention head, respectively.

Layer normalization [28] is incorporated to stabilize and expedite the training process by normalizing layer inputs. For an input vector $\mathbf{h} = [h_1, h_2, \ldots, h_d]$ with d features, the normalized output $\hat{\mathbf{h}} = [\hat{h}_1, \hat{h}_2, \ldots, \hat{h}_d]$ is computed as:

$$\hat{h}_i = \frac{h_i - \mu}{\sqrt{\sigma^2 + \epsilon}} \tag{8}$$

where $\mu = \frac{1}{d}\sum_{i=1}^{d} h_i$ represents the mean, $\sigma^2 = \frac{1}{d}\sum_{i=1}^{d}(h_i - \mu)^2$ denotes the variance, and ϵ is a small constant ensuring numerical stability. The final output y is obtained through the application of learnable scaling parameter γ and bias term β:

$$y_i = \gamma \hat{h}_i + \beta \tag{9}$$

3.3 The FGAT Framework

As illustrated in Fig. 1, the FGAT framework operates through a systematic process that begins with dynamic negative edge selection during each training epoch, utilizing the given adjacency matrix. These dynamically selected negative edges, along with the existing positive edges, are subsequently processed by the FGAT layer in conjunction with their corresponding node embeddings. To effectively capture long-range dependencies within the graph structure, multiple FGAT layers are cascaded, with residual connections implemented to enhance training stability and information flow. Following the iterative processing through these layers, we obtain updated node representations $H = \{h_1, h_2, \ldots, h_N\}$. The probability of link existence between any pair of nodes x and y is then computed as:

$$P_r^{link}(x, y) = \text{Sigmoid}(h_x h_y^T) \tag{10}$$

The framework's effectiveness stems from two key components: the fuzzy negative sampling technique, which efficiently identifies and selects high-quality negative edges, and the FGAT layer architecture, which performs iterative neighbor information aggregation. The empirical validation of this framework's performance is documented in the experiments section, demonstrating its effectiveness in link prediction tasks.

Table 1. Experiment Results

Methods	Ca-netscience				Ca-sandi-auths			
	Precision	Recall	F1	ROC	Precision	Recall	F1	ROC
MLP	0.4931	0.5879	0.5363	0.5044	0.5581	**1.0000**	0.7164	0.6181
GCN	0.5903	0.7363	0.6553	0.7078	0.5417	0.5417	0.5417	0.5304
GraphSAGE	0.5502	**0.8132**	0.6563	0.6277	0.4651	0.8333	0.5970	0.5885
GAT	0.6034	0.7692	**0.6763**	0.6916	0.6053	0.9583	0.7419	**0.7240**
FGAT	**0.6667**	0.6593	0.6630	**0.7422**	**0.6216**	0.9583	**0.7541**	0.7170

4 Experiments

We conduct comparative evaluations of FGAT against several state-of-the-art baselines using two research collaboration network datasets. The experimental results demonstrate the superior performance of FGAT. In this section, we present detailed information about the datasets, experimental settings, evaluation metrics, and analysis of results.

4.1 Datasets

Our evaluation utilizes two research collaboration network datasets, summarized in Table 2.

Table 2. Datasets Summary

	Ca-netscience	Ca-sandi-auths
#Nodes	379	86
#Edges	914	124
#AvgDegree	2.4	1.4
Directed	TRUE	TRUE
%Training Edges	0.7	0.7
%Validation Edges	0.1	0.1
%Testing Edges	0.2	0.2

The Ca-netscience dataset comprises 379 nodes and 914 edges, with an average node degree of 2.4. In contrast, Ca-sandi-auths exhibits a more sparse structure with fewer nodes, edges, and a lower average node degree. Both datasets are directed networks. For experimental purposes, we employ a 70-10-20 split ratio, where 70% of the data is used for training, 10% for validation (early stopping), and 20% for testing.

4.2 Experiment Settings and Evaluation Metrics

We benchmark FGAT against several prominent baseline models: MLP, GCN [3], GraphSAGE [1], and GAT [2]. All baseline models maintain their default parameter configurations. For FGAT implementation, we configure the embedding dimension to 128 and employ a stack of 4 FGAT convolution layers. The dataset partitioning follows the aforementioned 0.7:0.1:0.2 ratio for training, validation, and testing, respectively.

To ensure a comprehensive performance assessment, we employ multiple evaluation metrics: Precision, Recall, F1 score, and ROC score. This diverse set of metrics provides a multifaceted evaluation, enabling thorough analysis of each model's capabilities across different performance aspects.

4.3 Results

The experimental results are presented in Table 1, yielding several significant observations:

- On the Ca-netscience dataset, MLP demonstrates the poorest performance, attributable to its inability to capture spatial information encoded in the adjacency matrix. GCN, GraphSAGE, and GAT exhibit comparable performance levels, with GraphSAGE achieving superior Recall scores and GAT excelling in F1 metrics. For the Ca-sandi-auths dataset, MLP achieves notable Recall but relatively inferior Precision, suggesting overfitting tendencies and limited generalization capability. GAT, leveraging its attention mechanism, achieves the highest ROC scores among baseline methods.

- FGAT outperforms baseline methods across both datasets in terms of the average of four evaluation metrics. Specifically, it demonstrates an average improvement of 7.11% across all metrics on Ca-netscience, and a more substantial 15.55% improvement on Ca-sandi-auths. This superior performance can be attributed to two key factors: the fuzzy negative sampling mechanism, which enables focused learning on error-prone edges, and the FGAT layer architecture, which provides robust message aggregation capabilities.

5 Conclusion

The proposed FGAT framework, combining FNS and a novel graph attention layer, significantly improves link prediction performance compared to existing methods. FNS effectively identifies informative negative edges by leveraging fuzzy rough sets, leading to more focused and efficient model training. The FGAT layer, integrating fuzzy set concepts, captures complex relationships in graph data, resulting in superior node representations for accurate link prediction. The paper's findings highlight the potential of fuzzy rough sets in advancing GNNs for link prediction tasks and pave the way for future research exploring fuzzy set theory in graph-based learning.

References

1. Hamilton, W., Ying, Z., Leskovec, J.: Inductive representation learning on large graphs. In: Advances in Neural Information Processing Systems, vol. 30 (2017)
2. Veličković, P., Cucurull, G., Casanova, A., Romero, A., Lio, P., Bengio, Y.: Graph attention networks. arXiv preprint arXiv:1710.10903 (2017)
3. Kipf, T.N., Welling, M.: Semi-supervised classification with graph convolutional networks. arXiv preprint arXiv:1609.02907 (2016)
4. Ying, Z., You, J., Morris, C., Ren, X., Hamilton, W., Leskovec, J.: Hierarchical graph representation learning with differentiable pooling. In: Advances in Neural Information Processing Systems, vol. 31 (2018)
5. Diehl, F.: Edge contraction pooling for graph neural networks. arXiv preprint arXiv:1905.10990 (2019)
6. Zhang, M., Chen, Y.: Link prediction based on graph neural networks. In: Advances in Neural Information Processing Systems, vol. 31 (2018)
7. Kipf, T.N., Welling, M.: Variational graph auto-encoders. arXiv preprint arXiv:1611.07308 (2016)
8. Xing, J., et al.: NetSight: graph attention based traffic forecasting in computer networks. arXiv preprint arXiv:2505.07034 (2025)
9. Dubois, D., Prade, H.: Rough fuzzy sets and fuzzy rough sets. Int. J. Gen Syst **17**(2–3), 191–209 (1990)

10. Jensen, R., Shen, Q.: Fuzzy-rough attribute reduction with application to web categorization. Fuzzy Sets Syst. **141**(3), 469–485 (2004)
11. Xing, J., Gao, C., Zhou, J.: Weighted fuzzy rough sets-based tri-training and its application to medical diagnosis. Appl. Soft Comput. **124**, 109025 (2022)
12. Gao, C., Zhou, J., Xing, J., Yue, X.: Parameterized maximum-entropy-based three-way approximate attribute reduction. Int. J. Approximate Reasoning **151**, 85–100 (2022)
13. Ye, J., Zhan, J., Ding, W., Fujita, H.: A novel fuzzy rough set model with fuzzy neighborhood operators. Inf. Sci. **544**, 266–297 (2021)
14. Ji, W., et al.: Fuzzy rough sets and fuzzy rough neural networks for feature selection: a review. Wiley Interdisc. Rev. Data Min. Knowl. Discovery **11**(3), e1402 (2021)
15. Lin, C., Chen, S., Yang, X., Li, C., Qu, C., Chen, Q.: Research and application of knowledge graph technology for intelligent question answering. In: 12th International Symposium on Parallel Architectures, Algorithms and Programming (PAAP), pp. 152–156. IEEE (2021)
16. Yang, X.: Research on automatic composition based on multiple machine learning models. In: 2021 3rd International Conference on Artificial Intelligence and Advanced Manufacture, pp. 1206–1209 (2021)
17. Yang, X., Guo, Z., Mai, Z.: Botnet detection based on machine learning. In: 2022 International Conference on Blockchain Technology and Information Security (ICBCTIS), pp. 213–217. IEEE (2022)
18. Cheng, Q., Qu, S., Lee, J.: Patch-aware vector quantized codebook learning for unsupervised visual defect detection. In: 2024 IEEE 36th International Conference on Tools with Artificial Intelligence (ICTAI), pp. 586-592. IEEE (2024)
19. Cheng, Q., Xing, J., Xue, C., Yang, X.: Unifying prediction and explanation in time-series transformers via shapley-based pretraining. arXiv preprint arXiv:2501.15070 (2025)
20. Cheng, Q., Qu, S., Lee, J.: 72-3: deep learning based visual defect detection in noisy and imbalanced data. In SID Symposium Digest of Technical Papers, vol. 53, no. 1, pp. 971–974 (2022)
21. Cheng, Q., Zhang, C., Shen, X.: Estimation of energy and time usage in 3D printing with multimodal neural network. In: 2022 4th International Conference on Frontiers Technology of Information and Computer (ICFTIC), pp. 900–903. IEEE (2022)
22. Xing, J., Xiao, Z., Wu, Y., Zhang, J., Xu, Z., Mai, Z.: Network traffic forecasting via fuzzy spatial-temporal fusion graph neural networks. In: 2024 11th International Conference on Soft Computing & Machine Intelligence (ISCMI), pp. 282–286. IEEE (2024)
23. Cheng, Q., Qu, S., Lee, J.: SHAPNN: shapley value regularized tabular neural network. arXiv preprint arXiv:2309.08799 (2023)
24. Xing, J., Shahzad, M.: A reinforcement learning framework for application-specific TCP congestion-control. arXiv preprint arXiv:2505.07042 (2025)
25. Zhao, S., et al.: Image retrieval based on blockchain. In: 2022 International Conference on Blockchain Technology and Information Security (ICBCTIS), pp. 210–212. IEEE (2022)
26. Yang, X., Zhan, Y., Iwasaki, Y., Shi, M., Tang, S., Iwata, H.: Balancing real-world interaction and VR immersion with AI vision robotic arm. In: 2023 IEEE International Conference on Mechatronics and Automation (ICMA), pp. 2051–2057. IEEE (2023)

27. Deng, X., Zhou, H., Yang, X., Ye, C.: Short-term traffic condition prediction based on multi-source data fusion. In: International Conference on Data Mining and Big Data, pp. 327–335. Springer (2021)
28. Xiong, R., et al.: On layer normalization in the transformer architecture. In: International Conference on Machine Learning, pp. 524–533. PMLR (2020)

Towards Reliable Explainable AI: A Novel Stability Metric for Trustworthy Interpretations

Talal Ashraf Butt(✉) and Muhammad Iqbal

Higher Colleges of Technology, Fujairah, UAE
{tbutt,miqbal1}@hct.ac.ae

Abstract. Explainable Artificial Intelligence (XAI) methods-such as LIME and SHAP-are becoming essential for clarifying the internal logic of modern machine learning models. Despite their growing popularity, these techniques often yield explanations that shift substantially under minor input perturbations, undermining user trust. To tackle this challenge, we introduce an explanation stability metric, a novel measure that evaluates the robustness of explanations against slight variations in the input data. Our metric is validated through extensive experimentation across four datasets (Heart Disease, Adult, Breast Cancer, Synthetic), three model families (Logistic Regression, Random Forest, XGBoost), and two leading explainers (LIME, SHAP). The results confirm that stability uncovers unique insights beyond standard metrics like fidelity and faithfulness. In particular, simpler models tend to produce more stable explanations, SHAP exhibits stronger robustness than LIME, and dataset characteristics play a pivotal role in explanation consistency. By spotlighting the significance of stable explanations, this work guides researchers and practitioners in developing and deploying more dependable XAI approaches. The proposed metric promotes ethically sound and trustworthy AI implementations, particularly in critical fields such as healthcare and finance, ensuring that explanations remain consistent, reliable, and deserving of user confidence.

Keywords: Explainable Artificial Intelligence · Explanation Stability · Model Interpretability · XAI · Machine Learning · AI Ethics

1 Introduction

As Artificial Intelligence (AI) systems increasingly shape crucial decisions in healthcare, finance, and law sectors, the urgency to provide transparent and trustworthy outcomes continues to intensify [2]. Explainable AI (XAI) techniques, including LIME (Local Interpretable Model-agnostic Explanations) [6] and SHAP (SHapley Additive exPlanation) [4], aim to shed light on the reasoning behind model predictions, thus enabling stakeholders to understand, verify, and ultimately have confidence in these complex computational tools.

A. Ito (Ed.): ICMSSP 2025, LNNS 1637, pp. 198–210, 2025.
https://doi.org/10.1007/978-3-032-05994-9_17

Despite the growing popularity of XAI techniques, such as LIME and SHAP, in elucidating complex model decisions, recent findings reveal that these methods frequently exhibit instability under minor input perturbations [8,9,11]. Even negligible changes in the feature values can produce substantially different feature importance rankings or contribution values. As a result, users may encounter inconsistent explanations for nearly identical inputs, raising serious doubts about the reliability and trustworthiness of these explanation methods.

Traditional metrics, such as fidelity, which measures how well the explanation approximates the predictions of the underlying model, or comprehensibility and sparsity, which focus on human interpretability, do not capture the robustness of the explanations to minor input perturbations. Although these metrics are valuable, they do not address whether explanations remain reliable when real-world data changes are inevitable [10].

In safety-critical or sensitive applications, even small changes in input data may occur due to measurement errors, sensor noise, or natural variations. Stakeholders must be assured that the explanations do not dramatically oscillate with these minor changes. The absence of a measure dedicated to evaluating stability leaves practitioners without a systematic way to assess and compare XAI methods on this crucial dimension.

Hence, there is a pressing need for a dedicated metric that quantifies the stability of explanations. This metric can help researchers identify methods that produce consistent insights and guide practitioners toward deployable solutions that foster trust and reliability in AI-assisted decision-making. To meet this need, we introduce the *explanation stability* metric, which quantitatively measures how sensitive explanations are to slight input variations, thereby complementing traditional XAI evaluation criteria.

We demonstrate that stability is a crucial dimension of XAI quality through comprehensive experiments spanning multiple datasets, model architectures, and explainers. The key contributions of the proposed explanation stability metric are:

- Quantitative Robustness Assessment: It quantitatively measures how stable and consistent explanations are under small perturbations to the input, filling a gap not covered by fidelity or faithfulness metrics.
- Model and Explainer Differentiation: The explanation stability metric reveals differences in robustness between various explainers (LIME, SHAP) and models (Logistic Regression, Random Forest, XGBoost), assisting practitioners in choosing more reliable XAI methods.
- Dataset Sensitivity Insights: It shows how dataset characteristics influence the stability of explanations, guiding the selection of models and explainers for particular data domains.
- Complementarity to Existing Metrics: Stability complements fidelity and faithfulness, highlighting that good fidelity or faithfulness does not necessarily imply stable explanations.
- Foundation for Trust and Ethics: By promoting stable and consistent explanations, the explanation stability metric supports more trustworthy deployments of AI systems in sensitive and high-stakes environments.

By systematically evaluating and comparing stability alongside traditional metrics, we highlight new insights that were previously masked. This work advances the methodology for assessing explainability and offers practitioners guidance in selecting models, explainers, and data preprocessing strategies that lead to reliable and consistent interpretations-ultimately bolstering trust in AI's role in critical decision-making environments. Our findings reveal that:

- Simpler models, such as Logistic Regression, produce more stable explanations, while complex models like XGBoost exhibit higher variability.
- SHAP generally provides more stable explanations than LIME, indicating that theoretical underpinnings can enhance explanation robustness.
- The chosen dataset influences explanation stability, emphasizing that context matters when evaluating and deploying XAI solutions.

2 Explanation Stability Metric

To rigorously capture explanation stability, we focus on consistency in feature importance values across perturbed inputs. Rather than relying on the internal model mechanics, our proposed metric directly measures how much the reported feature attributions (e.g., weights, importance, or Shapley values) change when the input is slightly altered.

Suppose $\mathbf{x}$ is the original instance to be explained by an explainer E. The proposed explanation stability metric S is derived as follows:

- Let $(w_1, \ldots, w_d) = E(\mathbf{x}, M)$ represent the baseline explanation produced by explainer E for model M.
- Let $\{\mathbf{x}'_i\}_{i=1}^n$ be a set of n perturbed instances of $\mathbf{x}$, and let $(w'_{i1}, \ldots, w'_{id}) = E(\mathbf{x}'_i, M)$ be the corresponding explanations.
- Define the feature set:

$$F = \{f : w_f \neq 0 \text{ or } w'_{if} \neq 0 \text{ for some } i\},$$

  i.e., the union of features that are assigned any importance in either the baseline or perturbed explanations.
- For each perturbed instance, we compute:

$$S_i = 1 - \frac{\sum_{f \in F} |w_f - w'_{if}|}{|F|}.$$

  Here, S_i measures how close the perturbed explanation is to the baseline. Higher S_i values indicate that the explanation remains consistent despite the perturbation.
- Finally, the explanation stability metric S is calculated as:

$$S = \frac{1}{n} \sum_{i=1}^{n} \left(1 - \frac{\sum_{f \in F} |w_f - w'_{if}|}{|F|}\right)$$

Since the explanation stability metric S focuses on the relationship between explanations, it does not require directly modifying or probing the model's sensitivity. Instead, it treats the explainer as a function that maps instances to feature attributions and measures how stable these attributions remain under slight input changes. The algorithm 1 presents the proposed stability metric in algorithmic form.

Input : Original instance $\mathbf{x}$;
Set of perturbed instances $\{\mathbf{x'}_i\}_{i=1}^n$;
Predictive model M;
Explainer E.
Output: Stability score S.
Step 1: Compute baseline explanation:
$(w_1, \ldots, w_d) \leftarrow E(\mathbf{x}, M)$.
Step 2: Compute perturbed explanations:
foreach $i = 1, \ldots, n$ **do**
| $(w'_{i1}, \ldots, w'_{id}) \leftarrow E(\mathbf{x'}_i, M)$.
end
Step 3: Determine feature set:
$F \leftarrow \{f : w_f \neq 0 \text{ or } w'_{if} \neq 0 \text{ for some } i\}$.
Step 4: Compute instance-level stability:
foreach $i = 1, \ldots, n$ **do**
| $S_i \leftarrow 1 - \frac{\sum_{f \in F} |w_f - w'_{if}|}{|F|}$.
end
Step 5: Aggregate Stability:
$S \leftarrow \frac{1}{n} \sum_{i=1}^n S_i$.
Step 6: Return:
return S.

Algorithm 1: Simplified Explanation Stability Calculation

This algorithmic framework ensures the metric is model-agnostic and can be applied uniformly to any explainer. Focusing on the explanation output's stability rather than the internal model state offers a straightforward, flexible tool for benchmark comparisons and guiding improvements in future XAI methods.

3 Experiments

3.1 Experimental Setup

We evaluate the proposed explanation stability metric using four binary classification datasets: UCI Heart Disease (HD) [3], UCI Adult (AD) [1], UCI Breast Cancer Wisconsin (BC) [12], and a synthetic dataset (SY). These datasets offer a range of complexity and feature distributions, enabling a comprehensive stability assessment under diverse conditions. Table 1 provides a summary of these datasets.

Table 1. Summary of the Datasets Used in the Experiments

Dataset	# of Samples	# of Features
UCI Heart Disease (HD)	303	13
UCI Adult (AD)	48,842	14
UCI Breast Cancer Wisconsin (BC)	569	30
Synthetic Dataset (SY)	1000	20

We evaluate three classification models of varying complexity:

- **Logistic Regression (LR):** A fundamental linear model that leverages a logistic function for class probability estimation [13]. It offers computational simplicity and strong interpretability in linearly separable scenarios.
- **Random Forest (RF):** An ensemble of decision trees that votes for a final classification prediction [7]. Aggregating multiple tree outcomes typically mitigates overfitting and enhances predictive accuracy.
- **XGBoost (XGB):** A powerful gradient boosting approach recognized for its exceptional scalability and performance [5]. It incrementally refines errors through successive trees, making it well-suited to complex datasets.

For explanations, we use two widely-adopted methods:

- **LIME (Local Interpretable Model-agnostic Explanations):** Generates interpretable, local approximations by fitting a simple model (e.g., linear) around the prediction of an individual instance, elucidating how input features contribute to the model's decision [6].
- **SHAP (SHapley Additive exPlanations):** Provides comprehensive feature attribution by computing Shapley values from cooperative game theory, ensuring a fair distribution of contribution across features, thus offering consistent and theoretically grounded insights [4].

We split each dataset into training and test sets with a 20% test size. We select 10 instances from the test set for each model and explainer combination. Each instance is perturbed 10 times by slightly adjusting one or more features according to a perturbation standard deviation of 0.05. We run each experiment with multiple random seeds to ensure robustness.

Table 2 summarizes the key parameters. This experimental design comprehensively evaluates how stability varies across datasets, models, and explainers and how it relates to conventional metrics like fidelity and faithfulness (Table 2).

3.2 Metrics Evaluated

To comprehensively assess the performance and reliability of the XAI methods, we evaluate three key metrics, each providing a distinct perspective on the quality of the generated explanations. These metrics are defined as follows:

Table 2. Experimental setup parameters.

Parameter	Value
Datasets	HD, AD, BC, SY
Models	LR, RF, XGB
Explainers	LIME, SHAP
Test Size	0.2
Instances to Explain	10
Number of Perturbations	10
Number of Features Explained	3 (selected based on importance)
Perturbation Std	0.05
Random Seeds	10

- **Fidelity:** Fidelity evaluates how closely the explanation reproduces the model's internal behavior, indicating whether the explanation can reliably predict the model's output for various inputs. High fidelity ensures that the explanation represents the model's decision-making process, providing users with an accurate reflection of the underlying mechanics. A low fidelity score indicates a significant divergence between the explanation and the model's actual behavior, highlighting potential discrepancies that could mislead users or obscure critical aspects of the model's operation.
- **Faithfulness:** Faithfulness assesses how well the explanation reflects the model's actual response when certain features are altered or removed. A faithful explanation accurately conveys the significance of each feature in shaping predictions. This metric gauges how well the explanation reflects the importance of features in influencing the model's output. A faithful explanation reliably correlates with the model's response to feature perturbations, reinforcing the interpretability and validity of the XAI method. In contrast, an unfaithful explanation may inaccurately represent the contributions of different features, leading to misinterpretations and potentially erroneous conclusions.
- **Stability (Proposed):** Stability captures the degree to which small perturbations in the input data affect the explanation. High stability indicates that an explanation remains consistent under minor variations, improving user trust. Conversely, low stability suggests that explanations are highly sensitive to slight input variations, potentially undermining their trustworthiness in practical scenarios. Our proposed stability metric addresses a critical gap in evaluating XAI techniques by explicitly quantifying this sensitivity, thus offering insights into the resilience of explanations under non-ideal conditions.

While fidelity and faithfulness focus on the accuracy and representational quality of explanations, stability highlights their consistency and resilience against input variations. This holistic evaluation framework enhances the understanding of XAI robustness and facilitates the development of more reliable and interpretable AI systems.

4 Results

In this section, we present and analyze the results obtained from our experiments. We highlight how the explanation stability metric differentiates models, explainers, and datasets, complementing traditional metrics.

4.1 Relationship of Stability with Faithfulness and Fidelity

To investigate how stability relates to other metrics, we present Fig. 1, which plots faithfulness and fidelity against stability. The scatter plot illustrates the relationship between mean fidelity (x-axis) and mean faithfulness (y-axis) for SHAP explanations across different models, with point colors representing normalized stability. A gradient color bar on the right maps stability values, transitioning from blue (low stability) to red (high stability). The plot reveals that all data points cluster tightly around a mean fidelity of zero, suggesting that SHAP explanations generally produce consistent fidelity scores across models and datasets. However, the vertical spread along the faithfulness axis indicates variability in how well these explanations align with model predictions when features are perturbed.

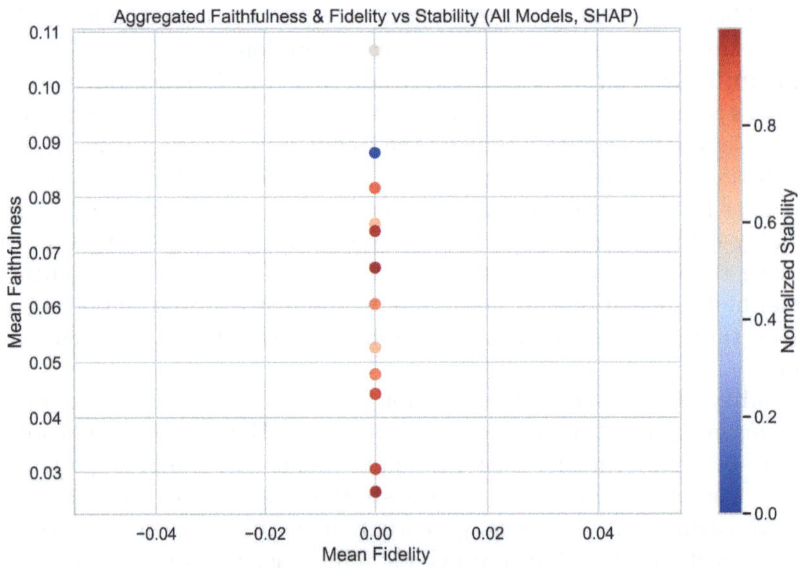

Fig. 1. Scatter plot illustrating the relationship between faithfulness, fidelity, and stability.

The color distribution highlights that higher stability (darker red) corresponds with mid-range faithfulness values, while lower stability (blue) aligns with outliers at the higher end of the faithfulness axis. This suggests a potential tradeoff between stability and faithfulness, where explanations that align more closely with model behavior might sacrifice robustness to small input changes. The plot underscores the importance of balancing these three metrics when evaluating the XAI methods, reinforcing that high fidelity alone does not guarantee stable or faithful explanations across different instances.

4.2 3D Visualization of Stability by Dataset and Model

Figure 2 provides a 3D visualization of mean stability scores. The 3D plot visualizes the mean stability scores of explanations generated by two popular XAI methods, LIME and SHAP, across different datasets and models. Stability is plotted along the z-axis, with datasets (AD, BC, HD, SY) on the x-axis and models (LR, RF, XGB) on the y-axis. The stability scores range between 0.93 and 1.00, providing insight into the robustness of the explanations across different configurations.

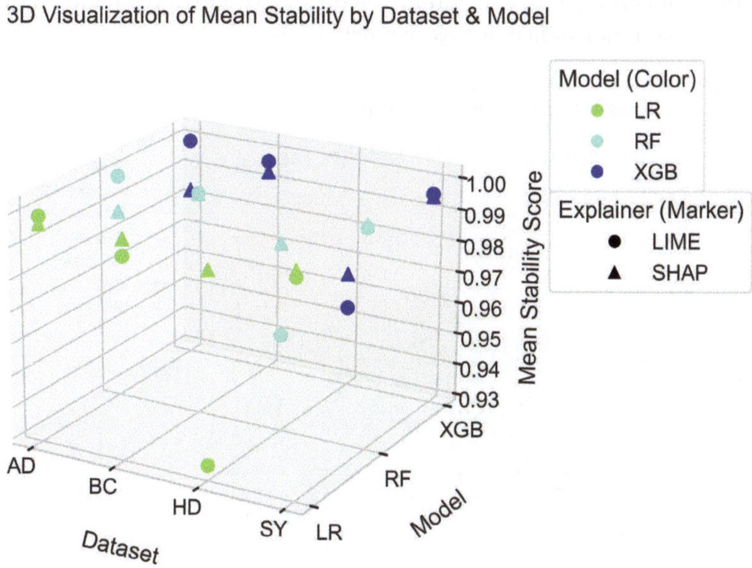

Fig. 2. 3D visualization of mean stability by dataset and model. Colors and markers indicate different explainers.

The analysis reveals that simpler models, such as Logistic Regression, exhibit consistently higher stability across all datasets, with both LIME and SHAP performing similarly well. In contrast, more complex models like XGBoost demonstrate more significant variability, with SHAP generally providing more stable explanations than LIME. The AD and BC datasets yield higher stability scores than the Synthetic dataset, which introduces more variability, particularly for Random Forest and XGBoost models. This suggests that model complexity and dataset characteristics significantly influence explanation stability, emphasizing the importance of explainer selection, particularly for non-linear ensemble models.

4.3 Single Instance Stability over Perturbations

To understand how stability changes with perturbation intensity, Fig. 3 plots single-instance stability as we increase the perturbation index. The graph illustrates the stability scores of three machine learning models, Logistic Regression, Random Forest, and XGBoost, evaluated using two explainability techniques, LIME and SHAP, across different perturbation indices. The x-axis represents the perturbation index, indicating the level or extent of perturbations applied to the input data. At the same time, the y-axis quantifies the stability score, a measure of consistency in the model explanations under these perturbations. The plot provides insight into how stable each model-explainer combination remains as the input data is perturbed. The shaded areas around each line signify the variability or uncertainty in the stability scores, with narrower bands indicating more consistent performance across perturbations.

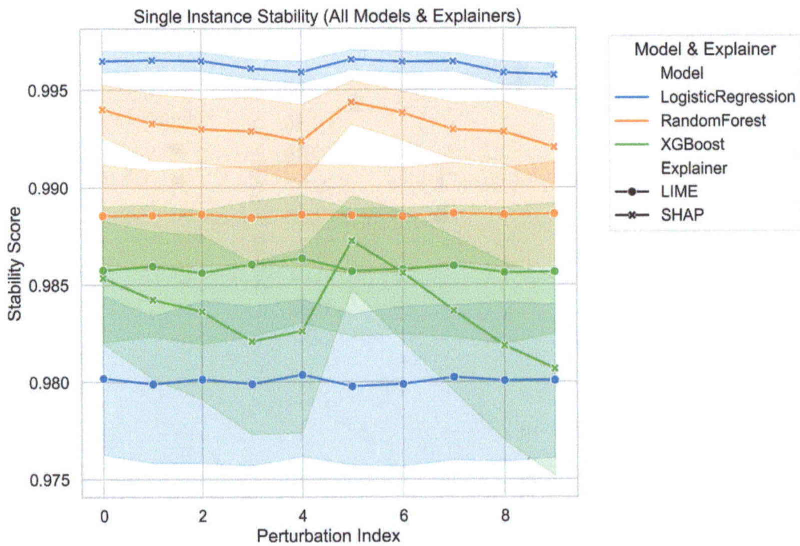

Fig. 3. Single-instance stability over perturbations for selected models and explainers.

The graph shows that Logistic Regression consistently achieves the highest stability scores, particularly when evaluated using SHAP, with minimal variance across perturbations. Random Forest also demonstrates relatively high stability, with SHAP slightly outperforming LIME. In contrast, XGBoost exhibits lower stability, and its performance degrades more noticeably at higher perturbation indices, especially with SHAP. This suggests that SHAP explanations for

XGBoost may be more sensitive to input perturbations than Random Forest and Logistic Regression. Overall, the graph highlights the varying robustness of explainability methods across different models, emphasizing the importance of selecting appropriate explainer techniques based on the underlying model's characteristics and sensitivity to perturbations.

4.4 Dataset-Specific Comparisons

Figure 4 visualizes the stability scores of the XGBoost model under the SHAP explainability technique across the datasets of adult, breast cancer, heart disease, and the synthetic dataset, as perturbations are introduced to the input data. The x-axis represents the perturbation index, reflecting incremental levels of perturbations, while the y-axis displays the stability score, quantifying the robustness of the model's SHAP explanations. Each line represents a different dataset, with shaded regions indicating the variability or confidence intervals for the stability score. The graph facilitates a comparative analysis of how the stability of SHAP explanations for XGBoost varies depending on the dataset under evaluation.

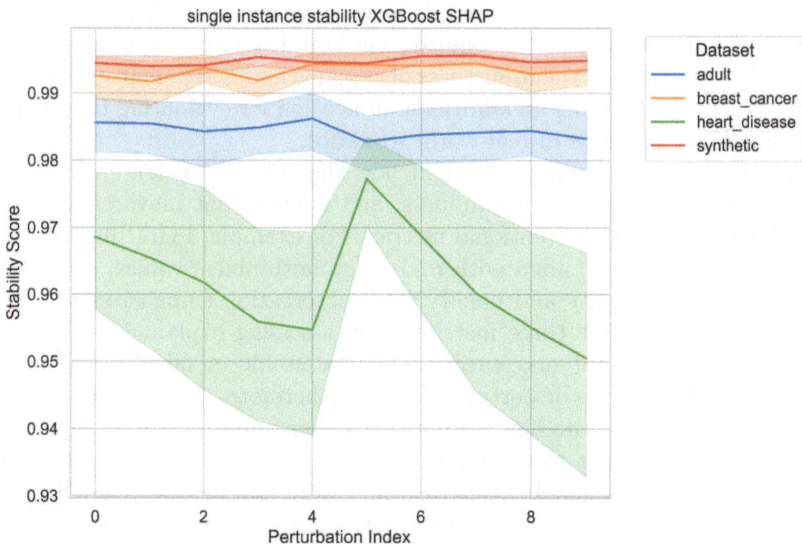

Fig. 4. Dataset-specific stability comparisons for a chosen dataset

The plot reveals that the synthetic and breast cancer datasets consistently exhibit the highest stability scores, with minimal fluctuation across perturbations. The adult dataset follows, demonstrating moderate stability with slightly larger confidence intervals. Conversely, the heart disease dataset exhibits the lowest stability, with a noticeable decline in stability scores as perturbations increase. The broader shaded area for the heart disease dataset suggests higher variability and sensitivity to input perturbations. This discrepancy in performance highlights the dataset-dependent nature of SHAP explanations when applied to XGBoost, underscoring the importance of dataset characteristics in determining explainability robustness.

The experimental results reveal that: (1) simpler models (e.g. LR) produce more stable explanations, suggesting that complexity complicates interpretability, (2) SHAP generally offers more stable explanations than LIME, indicating that certain theoretical foundations help maintain robustness under perturbations, and (3) stability adds valuable insights beyond fidelity and faithfulness. While a model might produce high-fidelity or faithful explanations, they may still vary wildly with small input changes, eroding user trust.

5 Discussion

This paper's key contribution is the explanation stability metric, a new evaluation tool designed to measure how consistently an explainer responds to slight variations in the input data. Traditional XAI metrics, such as faithfulness and fidelity, focus on how well an explanation mirrors a model's inner workings or predictive behavior. However, they do not assess whether those explanations remain stable when inputs shift slightly. The stability metric addresses this gap by explicitly quantifying the robustness of explanations across multiple perturbation levels, thereby offering an additional dimension for evaluating the reliability of explainability methods in real-world scenarios.

Our experimental findings highlight how the stability metric unearths relationships between stability, faithfulness, and fidelity that may remain hidden when relying solely on conventional metrics. For example, Fig. 1 illustrates that higher stability tends to align with mid-range faithfulness values, suggesting a tradeoff between stability and fidelity. This observation emphasizes the importance of balancing these three metrics when assessing explainability methods, as high fidelity does not necessarily guarantee stable or faithful explanations across different instances or perturbations. Furthermore, the 3D visualization in Fig. 2 provides insights into how stability varies across different models, datasets, and explainers, showing that simpler models (e.g., Logistic Regression) tend to exhibit higher stability. In contrast, more complex models (e.g., XGBoost) demonstrate more significant variability, particularly when using certain explainers.

In addition to providing insights into the stability of model explanations across different models and explainers, the stability metric also sheds light on the sensitivity of explanations to input perturbations. Figure 3 demonstrates how stability scores evolve as perturbations increase, revealing that simpler models maintain stable explanations while more complex models like XGBoost experience significant degradation in stability. This suggests that explainers for complex models may need to be more robust to perturbations, highlighting the importance of selecting appropriate explainers based on the model's complexity.

The analysis further emphasizes that dataset characteristics play a crucial role in stability, as shown in Fig. 4, where datasets like synthetic and breast cancer yield higher stability scores compared to the heart disease dataset, which exhibits lower stability under perturbations. The stability metric thus enables a more nuanced understanding of how datasets and models influence the reliability of explanations. Overall, the proposed stability metric offers a comprehensive approach to evaluating explainability techniques, complementing traditional metrics and providing a more complete picture of model transparency, robustness, and real-world applicability.

In real-world scenarios, choosing models and explainers that yield stable explanations can enhance confidence in AI-assisted decisions, particularly in critical domains like healthcare or finance. The observed dataset effects suggest that researchers should consider data properties when selecting or designing XAI techniques. Overall, the stability metric guides us towards more reliable, trustworthy, and context-aware explainability, encouraging future research on stable and consistent interpretability methods.

6 Conclusion

This paper introduced stability metric as a new dimension for evaluating XAI methods. Our experimental results confirm that stability is a critical yet previously overlooked property of explanations. While existing metrics (fidelity and faithfulness) capture certain aspects of explanation quality, they fail to address how consistently these explanations hold under slight perturbations to the input.

We uncovered patterns that would otherwise remain hidden by applying the stability metric to various datasets, models, and explainer methods. Simpler models produce more stable explanations, SHAP is generally more robust than LIME, and dataset characteristics significantly impact stability. These insights equip practitioners with actionable guidance for selecting model-explainer combinations that yield more reliable and trustworthy explanations.

While preliminary results suggest that our stability metric effectively captures pivotal properties of LIME and SHAP explanations, future work must examine its broader applicability to other techniques, such as counterfactual explanations and integrated gradients. Future research will also investigate alternative formulations of the stability metric, extend analyses to additional XAI methods and model architectures, and account for domain-specific nuances. We envision that our stability metric will ultimately encourage the development of next-generation XAI approaches, propelling the field toward explanations that remain consistently reliable under real-world conditions while retaining their informativeness and accuracy.

References

1. Becker, B., Kohavi, R.: Adult. UCI Machine Learning Repository (1996). https://doi.org/10.24432/C5XW20
2. Crook, B., Schlüter, M., Speith, T.: Revisiting the performance-explainability trade-off in explainable artificial intelligence (xai). arXiv:2307.14239 (2023)
3. Janosi, A., Steinbrunn, W., Pfisterer, M., Detrano, R.: Heart Disease. UCI Machine Learning Repository (1989). https://doi.org/10.24432/C52P4X
4. Lundberg, S.M., Lee, S.I.: A unified approach to interpreting model predictions. In: NIPS 2017, Proceedings of the 31st International Conference on Neural Information Processing Systems, pp. 4768–4777. Curran Associates Inc., Red Hook, NY, USA (2017)
5. Osman, A.I.A., Ahmed, A.N., Chow, M.F., Huang, Y.F., El-Shafie, A.: Extreme gradient boosting (xgboost) model to predict the groundwater levels in selangor malaysia. Ain Shams Eng. J. **12**(2), 1545–1556 (2021)
6. Ribeiro, M.T., Singh, S., Guestrin, C.: "why should i trust you?": explaining the predictions of any classifier. In: KDD 2016, Proceedings of the 22nd ACM SIGKDD International Conference on Knowledge Discovery and Data Mining, pp. 1135–1144. Association for Computing Machinery, New York, NY, USA (2016). https://doi.org/10.1145/2939672.2939778
7. Shaik, A.B., Srinivasan, S.: A brief survey on random forest ensembles in classification model. In: International Conference on Innovative Computing and Communications: Proceedings of ICICC 2018, vol. 2, pp. 253–260. Springer (2019)
8. Stevens, A., Smedt, J.: Explainability in process outcome prediction: guidelines to obtain interpretable and faithful models. Eur. J. Oper. Res. **317**(2), 317–329 (2024). https://doi.org/10.1016/j.ejor.2023.09.010
9. Sun, A., Sudjianto, A., Fargo, W., Zhang, A.: Evaluating perturbation-based post-hoc explainer fidelity using inherently interpretable models. https://adamsunn.github.io/assets/other/fidelity.pdf
10. Velmurugan, M., Ouyang, C., Moreira, C., Sindhgatta, R.: Evaluating fidelity of explainable methods for predictive process analytics. In: Nurcan, S., Korthaus, A. (eds.) Intelligent Information Systems, pp. 64–72. Springer International Publishing, Cham (2021)
11. Vimbi, V., Shaffi, N., Mahmud, M.: Interpreting artificial intelligence models: a systematic review on the application of lime and shap in alzheimer's disease detection. Brain Inform. **11**(10), 1–29 (2024). https://doi.org/10.1186/s40708-024-00222-1
12. Wolberg, W., Mangasarian, O., Street, N., Street, W.: Breast Cancer Wisconsin (Diagnostic). UCI Machine Learning Repository (1993). https://doi.org/10.24432/C5DW2B
13. Zou, X., Hu, Y., Tian, Z., Shen, K.: Logistic regression model optimization and case analysis. In: 2019 IEEE 7th International Conference on Computer Science and Network Technology (ICCSNT), pp. 135–139. IEEE (2019)

Author Index

A. Ito (Ed.): ICMSSP 2025, LNNS 1637, pp. 211–212, 2025.
https://doi.org/10.1007/978-3-032-05994-9

MIX
Papier aus verantwortungsvollen Quellen
Paper from responsible sources
FSC® C105338

If you have any concerns about our products,
you can contact us on
ProductSafety@springernature.com

In case Publisher is established outside the EU,
the EU authorized representative is:
Springer Nature Customer Service Center GmbH
Europaplatz 3, 69115 Heidelberg, Germany

Printed by Libri Plureos GmbH
in Hamburg, Germany